Ricardas V. Ralys

Experimental methods for measurement of vaporization enthalpies of ILs

Ricardas V. Ralys

Experimental methods for measurement of vaporization enthalpies of ILs

TGA and DSC - development and application

Südwestdeutscher Verlag für Hochschulschriften

Impressum/Imprint (nur für Deutschland/only for Germany)
Bibliografische Information der Deutschen Nationalbibliothek: Die Deutsche Nationalbibliothek verzeichnet diese Publikation in der Deutschen Nationalbibliografie; detaillierte bibliografische Daten sind im Internet über http://dnb.d-nb.de abrufbar.

Alle in diesem Buch genannten Marken und Produktnamen unterliegen warenzeichen-, marken- oder patentrechtlichem Schutz bzw. sind Warenzeichen oder eingetragene Warenzeichen der jeweiligen Inhaber. Die Wiedergabe von Marken, Produktnamen, Gebrauchsnamen, Handelsnamen, Warenbezeichnungen u.s.w. in diesem Werk berechtigt auch ohne besondere Kennzeichnung nicht zu der Annahme, dass solche Namen im Sinne der Warenzeichen- und Markenschutzgesetzgebung als frei zu betrachten wären und daher von jedermann benutzt werden dürften.

Coverbild: www.ingimage.com

Verlag: Südwestdeutscher Verlag für Hochschulschriften GmbH & Co. KG
Heinrich-Böcking-Str. 6-8, 66121 Saarbrücken, Deutschland
Telefon +49 681 37 20 271-1, Telefax +49 681 37 20 271-0
Email: info@svh-verlag.de

Approved by: Rostock, Universität Rostock, Diss., 2011

Herstellung in Deutschland:
Schaltungsdienst Lange o.H.G., Berlin
Books on Demand GmbH, Norderstedt
Reha GmbH, Saarbrücken
Amazon Distribution GmbH, Leipzig
ISBN: 978-3-8381-2912-9

Imprint (only for USA, GB)
Bibliographic information published by the Deutsche Nationalbibliothek: The Deutsche Nationalbibliothek lists this publication in the Deutsche Nationalbibliografie; detailed bibliographic data are available in the Internet at http://dnb.d-nb.de.

Any brand names and product names mentioned in this book are subject to trademark, brand or patent protection and are trademarks or registered trademarks of their respective holders. The use of brand names, product names, common names, trade names, product descriptions etc. even without a particular marking in this works is in no way to be construed to mean that such names may be regarded as unrestricted in respect of trademark and brand protection legislation and could thus be used by anyone.

Cover image: www.ingimage.com

Publisher: Südwestdeutscher Verlag für Hochschulschriften GmbH & Co. KG
Heinrich-Böcking-Str. 6-8, 66121 Saarbrücken, Germany
Phone +49 681 37 20 271-1, Fax +49 681 37 20 271-0
Email: info@svh-verlag.de

Printed in the U.S.A.
Printed in the U.K. by (see last page)
ISBN: 978-3-8381-2912-9

Copyright © 2011 by the author and Südwestdeutscher Verlag für Hochschulschriften GmbH & Co. KG and licensors
All rights reserved. Saarbrücken 2011

Contents

1 Introduction 3
2 Methods of determination of pressures of low volatile compounds 7
   2.1 Classification of methods 7
   2.2 Static methods 8
   2.3 Boiling point method (ebulliometry) 9
   2.4 Transpiration method 9
   2.5 Langmuir method 11
   2.6 Knudsen effusion method 13
   2.7 Chromatographic method 15
   2.8 Calorimetric methods 16
3 Development of the screening thermogravimetric method for a quick appraisal of enthalpies of vaporization of ILs 17
   3.1 State of the Art 17
   3.2 Experimental section 26
      3.2.1 Materials and chemicals 26
      3.2.2 TGA — Measurements of Vaporization Enthalpy 27
         3.2.2.1 Mass uptake 27
         3.2.2.2 Duration of the isothermal step 29
         3.2.2.3 Temperature range 30
         3.2.2.4 Purge gas flow rate 31
         3.2.2.5 Summary of the optimal conditions 32
   3.3 Results 32
      3.3.1 Test-measurements of vaporization enthalpies of the reference Molecular Liquids ... 32
      3.3.2 Data treatment of the TGA-measurements of molecular liquids 33
      3.3.3 Test-measurements of vaporization enthalpies of Ionic Liquids 35
      3.3.4 Data treatment of the TGA-measurements of ionic liquids 36
   3.4 Discussion 38
      3.4.1 TGA studies of l-alkyl-3-methylimidazolium bis (trifluoromethane sulfonyl) imides .. 39
      3.4.2 Validation of TGA vaporization enthalpies l-alkyl-3-methylimidazolium bis (trifluorornethanesulfonyl) imides with Langmuir Quartz Microbalance Method 45
      3.4.3 TGA studies of l-alkyl-3-methylimidazolium halides 47
      3.4.4 Determination of the Vaporization Enthalpy of $[C_2mim]_2[Co(NCS)_4]$ 49
      3.4.5 TGA – outlook and conclusions 52
4 Introduction to reaction differential scanning calorimetry 53
   4.1 Experimental section 54
      4.1.1 Materials and chemicals 54

- 4.1.2 *Differential Scanning Calorimetry — Measurements of Reaction Enthalpy* 54
  - 4.1.2.1 In solvent or neat? 57
  - 4.1.2.2 Mole Ratio of Reagents 58
  - 4.1.2.3 Prove of completeness of the IL synthesis reaction 60
  - 4.1.2.4 Summary of the experimental conditions 60
  - 4.1.2.5 Temperature adjustment of the measured enthalpy of reaction to 298.15K 61
- 4.1.3 *Computations* 62
- 4.2 Results and discussion 62
  - 4.2.1 *State of the Art* 62
  - 4.2.2 *Reaction MeIm + Di-Ethylsulfate* 64
  - 4.2.3 *Molar Enthalpies of Formation of ILs from DSC-Measurements* 66
  - 4.2.4 *Indirect Determination of Molar Enthalpies of Vaporization of ILs* 68
  - 4.2.5 *Vaporization Enthalpy from Enthalpies of Formation* $\Delta_f H_{liq}^0 (IL)$ 69
  - 4.2.6 *Vaporization Enthalpy from DSC Enthalpies of Reaction,* $\Delta_r H_m^0$ 70
- 4.3 Structure — property relationships for thermodynamic properties of ionic liquids. Correlation with the number of C-atoms in the cation alkyl chain 73
  - 4.3.1 *Enthalpy of IL synthesis reaction,* $\Delta_r H_m^0$ 74
  - 4.3.2 *Enthalpies of vaporization of IL,* $\Delta_{vap}H_{298}$ 77
  - 4.3.3 *Enthalpies of formation of ILs in the liquid state,* $\Delta_f H_{m,liq}^0$ 80
  - 4.3.4 *Enthalpies of formation of ILs in the gaseous state,* $\Delta_f H_{m,g,298}^0$ 81
- 4.4 Calculation of the Gas Phase Enthalpies of Reaction and Formation. Quantum Chemical Calculations 83
  - 4.4.1 *Enthalpy of formation* $\Delta_f H_{m,g,298}^0$ *- improvement of the atomization procedure* 84
  - 4.4.2 *Quantum chemical calculations for degree of dissociation of the ion pair* 89
  - 4.4.3 *Quantum chemical calculations for degree of conversion of the IL synthesis reaction* 90
  - 4.4.4 *Enthalpy of the IL synthesis reaction. Composite Methods* 91
  - 4.5 DSC - conclusion 91
5 Outlook and conclusion of the work 92
References 95
Acknowledgments 104

1   Introduction

Ionic liquids (ILs) are the salts with the melting point being conventionally below 100°C. Due to such low melting temperature they are liquid under the room conditions in the most cases (so called room temperature ILs – RTILs). In contrast to molecular solvents ILs consist entirely of cations and anions in liquid state, and they can be considered as solvents.

ILs are formed mostly of an organic aliphatic or heterocyclic cation and organic or inorganic anion. Since ILs are prepared by coupling relatively large ions, it is possible to create novel functionalities by changing the structure of component ions [1]. Thus, properties of ILs may be varied in different ways. One way is to select suitable cation. Possible cations are as follows: 1,3-di - alkylimidazolium, 1 -alkylpyridinium, 1,1-di - alkylpiperidinium, 1,1-di - alkylpyrrolidinium, tetra - alky-lammonium, tetra - alkylphosphonium, etc. Variation of alkyl chain length may also be used for modification of cation. Another way of properties adjustment is to choose an appropriate anion. Anions usually used for ILs are halogenes, alkylsulfonates, alkyl-sulfates, tetrafluoroborate, hexafluoroborate, dicianamide, nitrate, trifluoromethansul-fonate (triflate, [OTf$^-$]), bis(trifluoromethylsulfonyl)imides (bistriflamide [NTf$_2^-$]), and tris (trifluoromethylsulfonyl) methanide ([CTf$_3^-$]). List of the ILs relevant to this work given in the table 1.1.

Table 1.1 The list of ILs investigated in this work

Cation \ Anion	[NTf$_2$]	[Cl]	[Br]	[EtSO$_4$]	[Co(SCN)$_4$]$^{2-}$
[C$_n$mim]	$n = 2 \div 16$	$n = 2 \div 8$	$n = 2 \div 8$	$n = 2$	–
[C$_2$mim]$_2^{2+}$	–	–	–	–	√

Consisting of large or/and asymmetric cation ILs have low melting point in contrast to inorganic salts. The reasons for that are low symmetry, weak intermolecular interactions and good distribution of charge. Low symmetry of the constituting ions prevents compact packing of molecules and consequently leads to low energy of the lattice and to the low melting point. The difference in crystalline lattice of inorganic solid salts and ILs is depicted in the fig. 1.1.

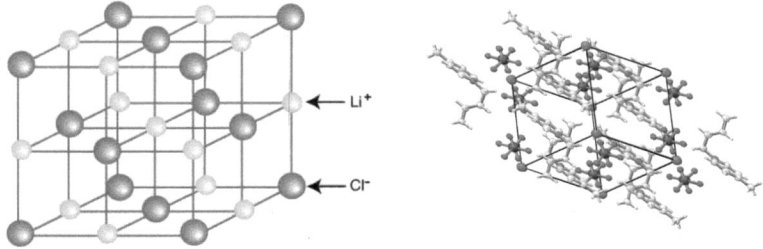

Figure 1.1: Comparison of crystalline lattice of inorganic salt and IL.

Selected properties, such as thermal stability and miscibility, mainly depend on the anion, while others, such as viscosity, surface tension and density, depend on the length of the alkyl chain in the cation and/or shape or symmetry [2]. The combination of the cited different anions and cations has expanded considerably the number of possible low temperature melting salts and more than two thousands ILs are known today. Beside their low melting point, they have a wide range of solubility, viscosity or density [3]. The extremely low vapor pressure of most ILs is the main reason that makes them useful in green chemistry. In chemical processes, they are easily recyclable and produce minimum pollution by volatile organic compounds if no organic solvent are used to recycle them. IL vapor pressure is most often non-measurable at room temperature. The extremely low volatility of ILs renders them little flammable so they could candidate to replace organic pollutant solvents [4]. ILs were quickly considered as benign or green solvents when full toxicity studies are not completed. It seems that many ILs have a significant ecotoxicity [5]. For example, the slow hydrolysis of the hexafluorophosphate anions released in water produces free toxic fluoride anions [6].

Low volatility has garnered them much recent attention as potential solvents to replace volatile organic solvents in a wide variety of chemical reaction, separation, and manufacturing processes. For example, ILs have been considered as solvents for reactions, as absorption media for gas separations, as the separating agent in extractive distillation, as heat transfer fluids, for processing biomass, and as the working fluid in a variety of electrochemical applications (batteries, solar cells, etc.) [7].

Understanding basic thermophysical properties of ILs is vital for design and evaluation. For instance, melting points, glass-transition temperatures, and thermal decomposition temperatures are needed to set the feasible temperature operating range for a particular fluid. Density (as a function of temperature) is needed for equipment sizing. Heat capacities are needed to estimate heating and cooling requirements as well as heat-storage capacity.

One of the main barriers for the development of ionic liquids for industrial applications is the scarce knowledge of their thermophysical properties [8], both for pure and mixed fluids, in the wide pressure and temperature ranges required for process design purposes. Some recent works have shown the most relevant aspects of thermophysical properties for ionic liquids [9]. The possible number of ionic liquids may be as large as 10^{18} (including binary and ternary ionic liquids) according to some authors [10], but a detailed analysis of the literature shows that basic properties have been measured only for a very limited number of ionic liquids [11].

It was claimed for a long time that one of the main characteristics of ionic liquids is that they are not volatile fluids, and thus, they would not show measurable vapor pressure, hence, they could not be distilled. Nevertheless, the possible volatility of ionic liquids was first explored by Rebelo et al. [12] and Paulechka et al. [13]. Although the pioneering work of Earle et al. [14] showed for the first time that ionic liquids could be distilled under reduced pressure at high temperatures, meanwhile at temperatures close to ambient conditions vapor pressure of ILs remains almost negligible. Moreover, the measurement of vapor-liquid equilibria properties was shown to be extremely difficult. The competing effects of fluids decomposition lead to the impossibility of measuring properties such as critical values. Likewise, the ionic character of these fluids leads to large vaporization enthalpies, $\Delta_{vap}H$, (in the 120-200 kJ mol^{-1} range [15, 16], but not as large as 300 kJ^{-1} mol as reported in some studies [12]), which are remarkably larger than for common organic fluids. A recent review by Esperanca et al. [17] analyzed in detail the available information on the volatility of aprotic ionic liquids, the main conclusions of the work were (i) at present, it may not be possible to measure the vapor pressures and/or the enthalpies of vaporization of any particular ionic liquid., and most methods present systematic errors related to the extreme experimental conditions of high temperatures that are not far from the decomposition temperatures of the ILs, and (ii) accurate measurement of enthalpies of vaporization for a sufficiently comprehensive set of ionic liquids will help researchers to refine force fields employed in molecular simulations as well as stringently test other theoretical models; these are absolutely essential tasks considering the enormous number of potentially available liquid salts.

In the present work development of express experimental methods (ThermoGravimetric Analysis – TGA and Differential Scanning Calorimetry – DSC) for studying of enthalpies of vaporization ILs was conducted. TGA is the technique allowing to derive enthalpy of vaporization directly from thermogravimetric data related to vapor pressure. In order to establish TGA for proper determination of $\Delta_{vap}H$ measurements with the reference substances were performed. With the use of these measurements experimental conditions indispensable for correct employment of TGA were found out.

Enthalpy of vaporization can be determined not only via vapor pressure or related to it values (like in TGA method), it can be obtained from enthalpies of formation of liquid and gaseous phases:

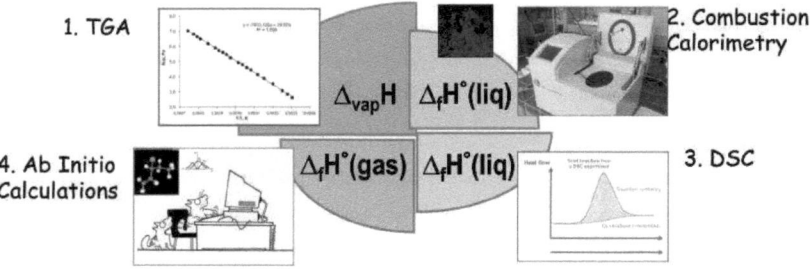

Figure 1.2: Thermodynamic properties of ILs - "upside-down" procedure.

$$\Delta_{vap}H = \Delta_f H_g^0 - \Delta_f H_l^0 \quad (1)$$

Enthalpy of formation in gaseous phase had been determined applying first principle calculations *(ab initio)*. Whereas enthalpy of formation in liquid phase was derived from calorimetric studies, using combustion calorimetry and DSC. From combustion calorimetry enthalpies of combustion of liquid ILs were measured and converted into the enthalpy of formation of liquids in standard state. In one turn, DSC is a technique which allows to measure enthalpies of reactions. From enthalpies of reaction of quaternization of ILs their enthalpies of formation in liquid phase can be evaluated. Comparison of enthalpies of formation of ILs in liquid phase derived both from combustion calorimetry and DSC was conducted to check mutual consistency of the data. The procedure when $\Delta_{vap}H$ is obtained from DSC (and combustion calorimetry) and *ab initio* we called "indirect" method, meanwhile the TGA method was called "direct" method in this work. Comparison of enthalpies of vaporization both from "direct" and "indirect" methods allows to check reliability of the data as well whether the IL under the study is stable under the conditions of TGA experiment. The described procedure is depicted schematically in the fig. 1.2.

Let us consider which principle methods and techniques for study of enthalpies of vaporization of low volatile substances are available today.

2 Methods of determination of pressures of low volatile compounds

Saturated vapor pressure is of a crucial importance in chemistry and chemical technology. This property is required is necessary for prediction of separation processes and distillation and reactors design. In addition such data allow to analyze the process of distribution of substances in environment. This analysis is very important for constructing new chemical plants. Moreover, the dependence of saturated vapor pressure on temperature gives the information about fundamental thermodynamic properties of a substance, which are related to its structure (e.g. enthalpy and entropy of vaporization).

2.1 Classification of methods

Experimental methods can be classified in different ways. One of possible approaches is to classify them according to construction of experimental set up of specific method. According to this approach it is possible to define the following techniques, which are broadly used for investigation of organic compounds [18]:

1. Static methods. In these methods closed and evacuated vessels are used under isothermal conditions.

2. Boiling point method (ebulliometry). In this group of methods the condition of boiling of substances is applied: $p^0 = P$, where p^0 is saturated vapor pressure, which should be measured, and P is the pressure in the vessel. When this condition is met the boiling of the liquid appears. The temperature of boiling corresponds to the saturated vapor pressure.

3. Transpiration method (or gas saturation method). In this method the mass of the vaporizing substance caught by a gas flow is measured. Then the obtained mass is converted to vapor pressure with the application of the equation of state of ideal gases.

4. Langmuir method and its modifications. The process of vaporization occurs from free surface and the relation between the rate of mass loss and the value of vapor pressure (Langmuir equation) is applied in this method.

5. Knudsen effusion method and its modifications. The process of vaporization occurs from the cell with a small orifice. The rate of effusion of the gas from the cell is related to the value of vapor pressure.

6. Chromatographic method. The experimental set up is the same as for classic chromatographic analysis. Generally speaking, the difference between such analysis and vapor pressure determination

is based on the application of the retention volume of the substance which is treated with the equation of state of ideal gas combined with Rault's law.

7. Calorimetric methods. This group of methods allows to measure only the enthalpy of vaporization which is derived from the temperature dependence of vapor pressure.

Let us consider all these methods in more detail.

2.2 Static methods

The main part of static method is a closed vessel filled with a sample, than the vessel is evacuated and the temperature is held constant within a time interval to reach the equilibrium between the condensed and vapor phases. The sample is carefully degassed prior to begin of the measurement. Pressure at equilibrium is determined with the aid of any measuring instrument. Many kinds of pressure gauges have been used, including mercury manometers, dead weight piston gauges, Bourdon tubes, and capacitance pressure transducers. In modern measurements of pressure high-precision manometers are commercially available (MKS, Ruska etc.), and different home-made gauges with sophisticated set up which are described earlier [19, 20] are out of use.

Despite its high sensitivity to volatile impurities and large time of experiment, static method provides us with the ability to measure pressures over a wide range of temperatures (200 K) with accurate temperature determination (±0.002 K) [21]. High-accuracy instruments are now commercial available and numerous of sophisticated homemade devices described in earlier compilations [22, 23, 24] became only of historical interest. Several effects, however, may limit the accuracy of static methods, especially for systems at low pressures or at extreme temperatures. A perfectly air tight apparatus is difficult to obtain. It is not easy to remove adsorption and desorption effects inside tubing connections. Impurities, decomposition products, dissolved or desorbed gases can easily mask vapor pressures of compound under study. The liquid and vapor phases of a sample may not be in equilibrium. Equipment, such as temperature and pressure gauges, may not be properly calibrated. If proper precautions are not taken, differential pressure indicator hysteresis or special transport phenomena such as thermal transpiration or molecular flow can obscure low-pressure results.

Since the sensitivity of static methods is limited by the pressure value of ~0.01 Pa, their application to ILs investigation is possible only at elevated temperature range, where the vapor pressure of ILs becomes prominent. This may cause errors in determination of $\Delta_{vap}H$ due to potential decomposition at higher temperatures. Also, large actual times of experiment may lead to degradation of the sample.

2.3 Boiling point method (ebulliometry)

The main part of any ebulliometric apparatus is ebulliometer. The general idea of the ebulliometer is similar to the one of an ordinary open boiler operating at atmospheric pressure. Since ebulliometry is a relatively old technique, a lot of experimental units have been developed. As accurate determination of the temperature of boiling is the key point of the measurement, it is necessary to provide smooth boiling and absence of overheating. Certain construction of the ebulliometer depends on how the measurement is carried out. Two principle measuring schemes are possible in this method [22]. The first one is a slow varying of external pressure under constant temperature and the second one is a slow varying of the temperature under constant pressure. In both cases it is necessary to detect the boiling point and corresponding pressure and temperature. The second scheme is possible in two main modifications: the modification based on Cotrell pump, circulating both liquid and vapor phases [23, 24, 25], and the modification with open reflux boiler without a Cotrell pump [26], with diverse facilities reducing overheating and bumping of the boiling liquid are elaborated.

Operation of ebulliometric equipment is possible in the pressures range of 2 kPa and higher. Ebulliometry is a method of very high reliability, when the conditions provide correctness of determination of the boiling point. For investigation of ILs ebulliometry is not suitable in most cases, since it requires boiling of the sample at pressures that are hardly possible to reach for this class of compounds.

2.4 Transpiration method

The method was proposed first by Regnault in 1845 [27]. The application of this method was restricted till the invention of automated chromatographic techniques in 1970s. During the long time method led to erroneous measurements due to insufficient saturation of carrier gas and invalid extrapolation of the data to zero rate of gas flow [28]. Despite all these difficulties this method has strong advantages, e.g. the influence of small amounts of volatile impurities is minimal, it is possible to carry out the measurements within relatively short time, besides, the method allows to change the atmosphere by switching the purge gas. The temperature range of measurements is close to ambient temperature and is very wide, vapor pressures can be measured from approximately 3 kPa and downwards[18].

The main part of the measurement system in this technique is saturator, where the investigated substance is held under controlled temperature. Through the saturator either inert or reactive gas is passed with sufficiently low rate to allow the equilibrium pressure of the substance to be established. Than the carrier gas is trapped by sorbents or cryogenic traps, vaporized substance is collected and its mass is determined.

Assuming that the purge gas is saturated sufficiently and the mixture obeys ideal gas behavior (under the common conditions of the method the assumption of ideality is valid) Dalton's law and ideal gas law are applicable. This gives the following expression for the vapor pressure:

$$p_i = \frac{m_i R T_a}{M_i V} \tag{2}$$

where p_i is vapor pressure of investigated substance, Pa; m_i is the mass of the substance evaporated, g; R is universal gas constant, $J\,mol^{-1}\,K^{-1}$; T_a is the temperature in the saturator, K; M_i is molar mass of the substance, $g \cdot mol^{-1}$; V - total volume of the gas passed through the saturator, m³, $V = V_{gas} + V_i$, $V_{gas} >> V_i$.

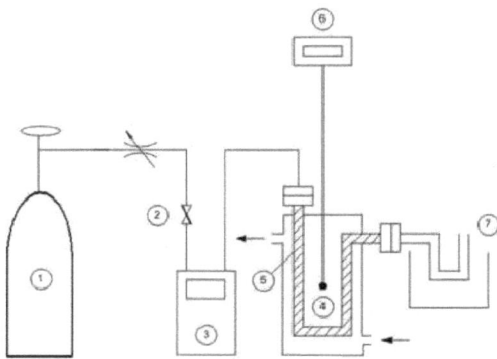

Figure 2.1: Schematic diagram of transpiration apparatus.

The above mentioned assumptions are met when the plateau in the plot of apparent pressure vs. flow rate of the carrier gas is reached. On the one hand, the flow rate of purge gas stream in the saturation tube should be not too low in order to avoid the transport of investigated substance due to diffusion. On the other hand the flow rate should be not too large in order to allow the purge gas stream to be saturated with a compound. Thus it is necessary to optimize the flow rate of the carrier gas stream through the saturator in order to keep the saturation of the transporting gas at each temperature under study. An approximate rates range, mentioned in the literature is about (1.5 to 7.0) $dm^3\,h^{-1}$.

Experimental set up consists of the saturator (glass tube - 5), which is immersed in a thermoliquid kept at constant temperature using a thermostat (4) with an accuracy of ±0.1 K. About 0.1-0.5 g of the investigated substance is mixed with glass beads or another support and placed in a full length of saturator. The temperature inside the saturator cell is also controlled by means of a platinum resistance thermometer (6) with an accuracy of ±0.1 K. The carrier gas from carrier gas cylinder (1) passes through the flow valve (2) and flow meter (3), enters the saturator, saturates with the substance along the tube and the saturated vapor condenses in a cold trap (7).

The transpiration experiment consists of two main parts. The first one is determination of the plateau in the plot of the vapor density versus the flow rate of the carrier gas at the mean temperature in the range of measurements. The second part is the determination of equilibrium vapor pressures at various temperatures from which the mean enthalpy of vaporization or sublimation is derived. In order to obtain pressure data the following experimental quantities are measured:

- mass of the sample transported by the carrier gas at the particular temperature. This quantity can be measured by chemical analysis of the condensate in the trap. Such approach provides the most accurate information, despite being quite time consuming. Several analytical methods can be applied for this kind of measurement, e.g. gas or high performance liquid chromatography, UV spectrometry, oxidation of condensate with IR analysis of produced carbon dioxide.
- flow rate of the carrier gas over the sample and the total time of sample condensation, from which the total volume of the gas swept over the sample is calculated; commercially available digital flow meters allow to reach high accuracy of such measurements.
- The temperature at which the transpiration experiment is performed.
- The ambient temperature at which the volume of the carrier gas is measured.

The accuracy of measurement of these quantities defines the accuracy of the results obtained.

In some recent works transpiration experiments combined with TG-method are described. In this case the loss of mass is measured directly by means of thermobalance and vapor pressure is evaluated [29, 30].

Despite this method requires a small amount of sample, and gives highly reproducible result (0.2--0.4 $kJ \cdot mol^{-1}$ for vaporization enthalpy), it has serious drawback particularly for investigation of ILs. Large actual time may be required to catch valuable amount of substance, and this may lead to thermal degradation of ILs.

2.5 Langmuir method

In 1910s Langmuir has found that during vaporization from open surface the rate of mass loss is proportional to the substance pressure and temperature according to the equation below:

$$\frac{dm}{dt} = \alpha SP \sqrt{\frac{M}{2\pi RT}} \qquad (3)$$

where $\frac{dm}{dt}$ is mass loss per unit time, a is coefficient of accommodation or Langmuir coefficient; S is the surface of sample; P is the pressure of the substance at the given temperature; M is molar mass of the vaporizing substance; R is universal gas constant; T is the absolute temperature at which the pressure is measured.

In classical experiments pressures of different metals were measured. In these measurements Langmuir heated wires of the substances being studied with an electric current in a *vacuum*. The temperature of the wire was measured with an optical pyrometer. The mass loss was obtained from radius of the wire as a function of time. Langmuir considered coefficient of accommodation as equal to unity in vacuum [31].

Price and Hawkins [32, 33, 34] proposed to use the equation (3) for the estimation of vapor pressures of low volatile substances from thermogravimetric *(TG)* data. They have found that, when the substance is vaporized not into vacuum but into flowing gas stream, accommodation coefficient is no longer equal to unity.

They rearranged the equation (3) into the following relationship:

$$P = \frac{\sqrt{2\pi R}}{\alpha} \frac{dm}{dt} \sqrt{\frac{T}{M}} \qquad (4)$$

With $k = \frac{\sqrt{2\pi R}}{\alpha}$ and $v = \frac{dm}{dt}\sqrt{\frac{T}{M}}$. The final expression for vapor pressure:

$$P = kv \qquad (5)$$

According to (5) the plot of P vs. v follows the same trend for a series of compounds with known vapor pressure, regardless of chemical structure, providing that the sample does not associate in the condensed or gas phase allowing the calibration constant k to be determined and thus the vapor pressures of unknown materials to be found [32, 34]. In the series of experiments by Dollimore and Alexander [35, 36, 37] the value for calibration constant k is found for a variety of low volatile substances and coefficient of accommodation was determined $a = 5.8 \cdot 10^{-5}$. Gückel et al. [38] observed vaporization rate being strongly affected by the flow rate. In contrast, Price and Hawkins [32] stated vaporization rate was not affected by small variations of purge gas rate.

Reference compounds are usually selected as those with well established vapor pressures. Experimental vapor pressures collected in the literature [39, 40] most often approximated by Antoine equation. Accuracy of the vapor pressures derived by TG directly depends on a quality of vapor pressures data sets of the chosen reference compounds. It is prudent to collect a set of reference substances with vapor pressure data measured in a possibly broad temperature range which is essentially close to the conditions of the TG experiment. In this context it is advisable to take some precautions by using the comprehensive compilation [39], which contains vapor-pressure results for all classes of organic compounds over a wide range of temperature. This compilation is willingly used by TG-community to extract the Antoine coefficients. However, the origin of the data presented in [39] is unclear; methods of measurements are unknown as well as errors of measurements and purities of compounds that is why use of Antoine coefficients listed there in each case should be considered as questionable and some other sources ought to find out for the sake of comparison.

Since there is scarce of data on vapor pressures of ILs for the purpose of reference, TGA can be used only for determination of enthalpies of vaporization. As a method of determination of $\Delta_{vap}H$, TGA has several advantages: it has short experimental times, a few tens of milligrams are required only, and commercial TG-equipment is easily accessible. Drawback of the method is experimental temperature range which can be close to decomposition onset.

2.6 Knudsen effusion method

This method is based on measurements of rate of vaporization through a small orifice from the space saturated with vapor [41] at constant temperature. For vacuum conditions the following relation between vapor pressure and rate of vaporization was derived by Knudsen [28]:

$$\frac{\Delta m}{t} = p \cdot S \cdot K \sqrt{\frac{M}{2\pi RT}} \qquad (6)$$

where Δm is the mass loss of the sample during the time t; S is the orifice surface; K is Clausing's factor, taking into account finite thickness of the orifice walls, this value is known from the table given by Dushman [42].

There are two ways to determine the loss of mass. The first one is weighing of effused vapor on the cold surface placed above the orifice with the further spectroscopic or chromatographic analysis of collected mass. A typical Knudsen effusion apparatus is presented in Figure 2.1.

Several modifications of Knudsen method exist, among them one can find the following:

1. **Torsion-effusion method.** The effusion cell is in the form of horizontal cylinder with two orifices disposed on opposite sides of the cylinder. The cell is suspended on the wire. The molecules leaving the cell through these orifices produce a force on the cylinder resulting in a torque of the wire. From the value of this torque the vapor pressure can be derived. The temperature is measured by means of thermocouple, placed in the cell equal to effusion cell which is based beneath it. In this method the time of 4-5 days is required to complete the measurement. The disadvantage of the method is inability to measure exactly the sample temperature.

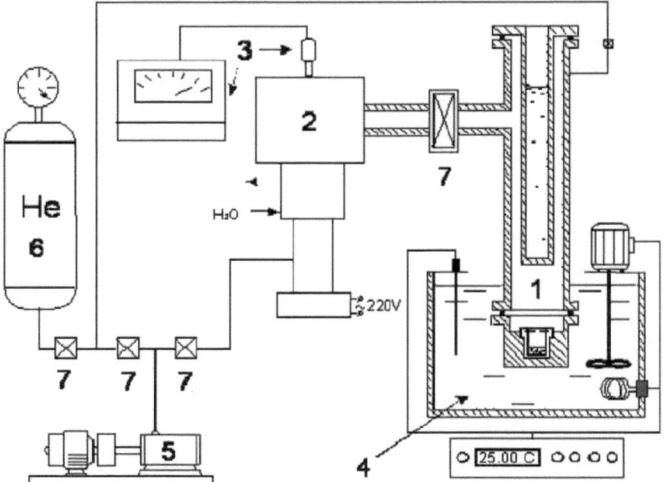

Figure 2.2: The scheme of the apparatus for vapor pressure measurement by Knudsen method: 1 - measuring block (effusion cell with a sample, vacuum connection, cold finger); 2 - diffusion oil pump; 3 - vacuum measuring gauge; 4 - thermostat; 5 - roughing-down pump; 6 - helium cylinder; 7 - bellows gates.

2. ***Isothermal Knudsen effusion method in the TG-type apparatus.*** In this technique the mass loss of the cell is recorded as a function of time in TG-type apparatus with high precision. In order to remove any volatile admixtures present in the sample at least 5% of the mass of the substance should be vaporized.

3. ***Non-isothermal Knudsen effusion method in the TG-type apparatus.***
This method was developed to satisfy the need to analyze the sample over a large range of temperatures during modest times. Unlike isothermal technique this method does not require the cell to reach the steady-state temperature.

4. ***Knudsen technique with a quartz crystal microbalance (QCM).*** The substance effusing the cell condenses on piezoelectric crystal and this leads to change in resonance frequency of the crystal. Since the mass left the cell is equal to the one condensed on the crystal, it is possible to measure the rate of effusion and evaluate vapor pressure. Though, there is a requirement of mass being uniformly distributed on the crystal.

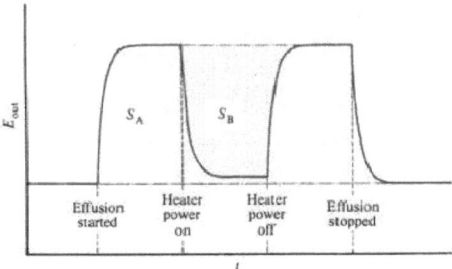

Figure 2.3: Schematic plot of output potential from the microcalorimeter against time for a sublimation calorimetric experiment. A calibration experiment is superposed on a sublimation experiment. S_A is the net area observed for the sublimation experiment. S_B is the area associated with the calibration.

5. Mass loss Knudsen technique with a heat-conducting calorimetry. Calorimetric determination of enthalpies of sublimation combined with Knudsen technique has been performed by use of Calvet-type heat-conduction calorimeter [43]. The record of an output signal of the calorimeter throughout an experiment is schematically depicted in figure . The enthalpy of sublimation at given temperature is calculated using the following equation: $\Delta_{sub} H = E_{calib} \dfrac{S_A + S_B}{S_B} \dfrac{M}{m}$

where E_{calib} is the electrical energy supplied to the calibration heater; m is the mass of the effused sample. Vapor pressure of sample is measured simultaneously and obtained by mass loss under isothermal conditions of Knudsen molecular effusion. The temperature behavior of p gives a mean value of $\Delta_{sub} H(T)$. Agreement between the mean values derived from calorimetric and from vapor pressure data provides a test of mutual consistency of the experimental results.

Knudsen method and especially Knudsen method with QCM is a suitable tool for accurate measurement of vapor pressures of ILs, since it has moderate actual times of experiment, high sensitivity of quartz crystal microbalance and moderate temperatures of experiment.

2.7 Chromatographic method

Chromatographic methods (GC) can be divided into two approaches: the dynamic, in which retention volume (time) of analyzed substance is measured, and the static (head-space) analysis, in which selective and sensitive gas analysis is performed.

A simple indirect static analysis for the determination of vapor pressures of pure substances was developed by Chickos[44]. A standard instrument consists of a sample chamber joined with ballast tank through a valve. Both compartments are kept at the same temperature, allowing the system to reach the equilibrium. The experimental value of the pressure of the investigated substance is obtained by measurement of the contents of the ballast tank and treatment of this quantity with the ideal gas law. It is necessary to take into account that this method gives underestimated values of vapor pressure, despite this, the enthalpy of vaporization remains reliable.

Dynamic method in one turn is presented by GC-correlations of retention times with vapor pressures of reference substances, and GC-correlations with net retention times. In both approaches the idea of linear dependence of the logarithm of vapor pressure on the reciprocal retention times is underlying. The difference between them consists of measurement of retention times of several substances with well-known pressure over desired temperature range in the first approach [45], meanwhile in the second approach the correlation for a wide range of substances from diverse chemical classes is used [46].

2.8 Calorimetric methods

This group of methods allow to measure the enthalpy of vaporization directly. Most of calorimeters are presented by adiabatic ones, where the heat absorbed during vaporization is exactly equal to the heat transferred from an external source. Another part of calorimetric determinations is performed by conduction calorimeters.

Let us consider the first group of methods, where adiabatic calorimeters are applied. At low pressures two approaches are employed, with vaporization into vacuum and with purging with carrier gas. Vaporization in vacuum is applicable even at pressures below 1 Pa, the upper limit is about 25 kPa. The range of pressures available for measurements in the systems with carrier gas is about 0.05 to 25 kPa. In both methods vaporization occurs in open space. At moderate pressures (5 ± 200 kPa) experiments are performed in closed systems, where vaporization and condensation parts exist.

Another possibility to measure the enthalpy of vaporization directly provides drop calorimetry method. The sample at room temperature is dropped in Calvet type calorimeter standing at enhanced temperature, providing rapid vaporization of the sample. The enthalpy of vaporization corresponds to the peak area, observed in the calorimeter [47].

3 Development of the screening thermogravimetric method for a quick appraisal of enthalpies of vaporization of ILs

3.1 State of the Art

Nowadays, the thermogravimetric analysis (TGA) is the well established experimental method for studies of a thermal behavior of a sample in a wide temperature range under controlled external conditions (purge gas, pressure or vacuum). The modern commercial TGA devices are fully automated and they provide highly accurate and reproducible results for the mass loss during heating of the sample. Connected to any spectroscopic unity (MS, IR, or UV) for analysis of the evolved gases, the TGA technique provides quantitative and qualitative information on thermal stability of the sample under study.

In this work we decided to extend fields of traditional applications of TGA and to elaborate the experimental conditions of a commercially available TGA for the fast determination of enthalpies of vaporization of extremely low volatile compounds. Unlike the thermal stability studies using the TGA, the temperature dependent mass loss of the sample, is intended to use for an express measurement of vaporization enthalpies of ionic liquids.

The principle of the TGA experiment is simple - 50-100 mg of IL-sample is placed into an open Pt-crucible (Fig. 3.1) [48] is heated with a certain rate up to elevated temperatures and the mass loss is continuously monitored at isothermal steps (see Fig. 3.1) . As a matter of fact it is hardly possible to separate this mass loss into contributions due to vaporization or due to decomposition, or even due to combination of these to processes. However, using the spectroscopic analysis of the residual IL in the Pt-crucible, the decomposition products could be easily detected or excluded. TGA method potentially has considerable advantages compared to the conventional methods for determining vapor pressures. Compared to conventional techniques for vapor pressure measurements [18, 49], TGA offers several advantages: a relatively small sample required for measurements, simple operation of the commercial setup, and short experimental times. As a rule TGA is used at ambient pressure with inert gas purge over measuring unity. In the latter case the sublimation or vaporization of the sample from open crucible is promoted by a gas purge over the phase contact surface and this procedure is essential for study of low volatile compounds.

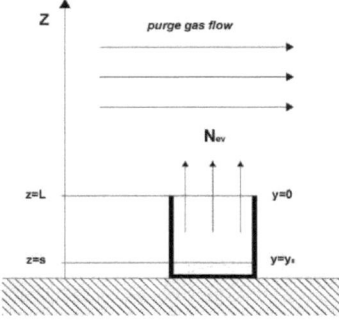 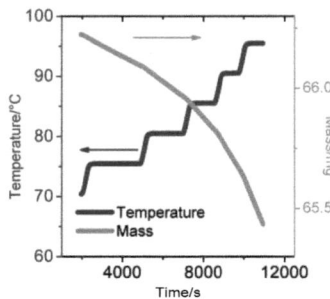

Figure 3.1: Measurements of $\Delta_{sub}H$ using TGA

Several procedures have been developed for investigation of vapor pressures and enthalpies of vaporization as well as kinetic parameters of vaporization process. Among them, isothermal measurements from the open surface (Langmuir experiments) in vacuum conditions, isothermal and non-isothermal zero-order kinetic measurements, non-isothermal diffusion controlled studies, Langmuir experiments in the presence of purge gas with the use of calibration compounds, and TG-based transpiration technique. Pioneering works by Langmuir were based on rigorous molecular-kinetic approach which led to derivation of Langmuir equation (3).

As a matter of fact, the Eq. (3) was developed for evaporation from open surface of solid compound into a vacuum. Langmuir considered coefficient of accommodation a as a value which should be equal to unity in such conditions. Langmuir equation has been successfully applied for measurements of vapor pressures of numerous inorganic solid compounds [38, 50, 51, 52, 53, 54, 55, 56, 57, 58, 32] under isothermal conditions with coefficient of vaporization equal to unity. These works had shown good agreement with the results available from other techniques, proving Langmuir method to be reliable for vapor pressure measurement of solid low volatile compounds. Development of the precise automatic electronic microbalances has open the way for broad applications of TG-based methods for vapor pressure measurements.

In the recent time Price and Hawkins [59, 32] formalized and adjusted a procedure for estimating the vapor pressure and vaporization enthalpies of low-volatility substances from isothermal thermogravimetric data using the Langmuir equation. Enthalpy of vaporization $\Delta_{vap}H$ at the average temperature of investigation was obtained from the slope of the Clausius-Clapeyron equation (7):

$$ln\left(\frac{dm}{dt}\sqrt{T}\right) = A - \frac{\Delta_{vap}H}{RT} \tag{7}$$

The main principle governing the TGA experiment was that evaporation and sublimation are zero-order processes. Hence, the rate of mass loss of a compound under isothermal conditions due to vaporization should be constant provided that the free surface area does not change [50].

From our knowledge, the use of thermogravimetry for study of vaporization process was first reported by Gückel et al. [38]. They measured the rates of mass loss of a group of pesticides using a self-recoding electronic balance in isothermal conditions. Gückel et al. [38] studied evaporation of a sample from the glass plate attached to the measuring arm of the balance. In order to prevent the stagnation of vapors near the surface a purged gas at fixed rates was applied. They observed a strong dependence of the evaporation rate on the rate of purge gas and suggested a calibration procedure at definite conditions using vapor pressures of a reference compound.

Elder [51] applied a commercially available TG device for sublimation studies of pharmaceutics into a nitrogen stream in an isothermal conditions. He pointed out the importance of proper choice of the temperature range for investigation. On the one hand, a noticeable mass uptake in an acceptable time is required and this restricts the lower limit of the temperature range. On the other hand, a thermal stability of compounds under study limits the upper temperature of the range.

Vaporization of series of dyes into a purge gas stream was investigated by Price et al. [33, 34] in non-isothermal conditions. They rearranged Langmuir equation for the simple data treatment (5). The calibration constant k was obtained from experiments with several reference substances with well known pressure under the fixed conditions. The TGA-procedure suggested by Price [59] is simple and fast, however the assumption that calibration constant derived from a few reference compound is valid for compounds of different structure and volatility needs to be worked out and tested carefully.

The TGA technique has been intensively used for studies of vapor pressures of pharmaceutical compounds [51, 52, 53], cosmetic ingredients [54, 55, 56, 57, 58], and dyes [32]. In the most studies the non-isothermal conditions of the TGA experiment have been applied.

For the successful measurements of vapor pressures using TGA the vaporization coefficient a in Eq. (1) have to be known. However, the practical and theoretical observation for the a is still in disarray. According to classic Langmuir's suggestion the vaporization coefficient a should be close to unity for low volatile compounds by vaporization in vacuum [31]. There are two classical approaches for estimating evaporation rates of pure liquids. The principle of detailed balance requires that under equilibrium conditions the flux of molecules that leave the surface (per unit area per unit time) be equal to the number of identical gaseous molecules that impact the surface, reduced by a sticking parameter (a < 1). It follows that the evaporation flux *(mol • s^{-1} • cm^{-2})* ejected into a vacuum, thus avoiding return by collisions with the ambient gas, be equal to the kinetic theory estimate of the collision rate of vapor molecules (per unit area, per unit time) at the equilibrium vapor pressure: $J = \alpha P / (2\pi pMRT)^{1/2}$, where P is the pressure, M the molecular weight. Thus, a remains a fitting parameter [60]. Another expression was proposed for the rate of evaporation in terms of a unimolecular dissociation of a species weakly bound to the surface layer, with an "activation energy" equal to the enthalpy of vaporization. Mortensen and Eyring [61] formulated this model in terms of "absolute reaction rate" theory, postulating a transmission coefficient to be much less than unity. Their "activated state" was defined at the instant when the evaporating species just separated from the surface layer. For the reverse step, condensation, the activation energy was assumed to be zero.

Jer Ru Maa [62] carried out calculations of the thermal gradient in liquid phase in order to analyze the reason why α < 1. Jer Ru Maa [62] investigated evaporation rates of water, isopropyl alcohol, carbon tetrachloride, and toluene. The experimental results agreed satisfactorily with the thermal gradient calculations, making the assumption that the evaporation coefficient is unity. This showed that there was little or no resistance to molecules crossing the vapor-liquid interface in addition to the natural resistance imposed by the gas laws. No significant difference in the behavior of evaporation due to the difference in molecular structure or chemical properties was observed.

Later, Price and Hawkins [32, 33, 34] recognized, that for vaporization into a flowing gas stream at atmospheric pressure rather than into a vacuum, the vaporization coefficient a might no longer be equal to unity. In order to assess the vaporization coefficient Pieterse and Focke [63] suggested a sophisticated correction for the diffusion of the sample into the gaseous phase. However, the introduction of plethora additional (but ill defined) parameters into the TGA treatment procedure makes it too complex for reliable estimation of the enthalpy of vaporization for extremely low-volatile compounds.

Nevertheless, in all determinations of the vapor pressure and the enthalpy of sublimation or vaporization carried out by TGA method condensation or accommodation coefficient a was taken to be constant in the temperature region of determination. The value of a in most of the cases depends on the used apparatus and vapor pressure determination technique, but not the compound and temperature interval.

In order to overcome estimation of the vaporization coefficient Dollimore and coworkers [52, 53, 54, 55, 56, 57, 58] also suggested using for TGA measurements a reference compound with the well established vapor pressures. According to their idea, the reference compound and the sample of interest were measured in the same TGA conditions. The vaporization coefficients for both compounds were supposed to be very similar and the corresponding vapor pressures were estimated.

However, in spite of the useful idea, the enthalpies of vaporization derived by Dollimore et al. [64] sometimes seem to be in error. For example, vaporization enthalpies of alkyl parabenes (esters of para-hydroxybenzoic acids) collected in Table 3.1 [64] are surprisingly very close. But it is well established, that the enthalpy of vaporization is linearly dependent on the number of carbon atoms in the alkyl chain with the increment for the CH_2 group of 4.6 to 5.0 $kJ \cdot mol^{-1}$ [65]. It means that enthalpy of vaporization of butyl parabene have to be at least 15 $kJ \cdot mol^{-1}$ higher than those for methyl parabene.

Table 3.1: Enthalpies of vaporization of series of parabenes

Substance	$\Delta_{vap} H$, $kJ\ mol^{-1}$
Methyl parabene	77.1
Ethyl parabene	75.0
Propyl parabene	79.1
Butyl parabene	76.9

Nevertheless, we have to acknowledge the efforts by Dollimore and co-workers to establish experimental conditions for reproducible TGA measurements.

Two important parameters were studied by Dollimore [66] in order to achieve the optimal conditions for the TGA experiment:

1. Variation the heating rate: The heating rate was scanned from 2 to 12 $K \cdot min^{-1}$. Purge gas (nitrogen) at the constant flow rate of 100 $mL \cdot min^{-1}$ was used in all experiments.

2. Variation of N_2 flow rate: the N_2 flow rate was changed from 50 to 1000 $mL \cdot min^{-1}$. The heating rate was maintained at 6 $K \cdot min^{-1}$ throughout the experiment.

For the reproducible determination of vapor pressure and enthalpy of vaporization, the N_2-flow rate 100 $mL \cdot min^{-1}$ and the heating rate between 8 to 10 $K \cdot min^{-1}$ has been recommended [66]. Later, Price [33] used the same flow-rate, but his heating rate was significantly lower (only 1 $K \cdot min^{-1}$).

Barotini and Cozzani [48] reported vaporization of substances with well established vapor pressures under isothermal conditions. They studied an influence of crucible and total pressure in the system with a focus on the diffusion processes. Using mass balance of vaporization process and Fick's law they developed the following equation:

$$N_{ev} = S_c \frac{DP}{RT(L-s)} ln\left(1 - \frac{p}{P}\right) \qquad (8)$$

where N_{ev} is molar evaporation rate, $mol \cdot s^{-1}$; S_c is surface available for mass transfer m^2; D is diffusion coefficient of the substance in the purge gas, $m2 \cdot s^{-1}$; P is total pressure in the system, assumed to be equal to that of purge gas, Pa; T is temperature of the measurement, K; L is height of the crucible, m; s is height of the substance layer in the crucible, m; p is vapor pressure of the substance, Pa.

In order to treat the equation (8) Barotini and Cozzani [48]. suggested to use additional equations, describing dependence of diffusion coefficient and vapor pressure on the temperature. According to Chapmen-Enskog model the dependence of the diffusion coefficient is defined in the following way:

$$D = k_0 T^{3/2} \qquad (9)$$

where k_0 - an empirical constant, defined from experimental data. To describe the vapor pressure dependence Antoine equation was used:

$$log_{10} p = A - \frac{B}{C+T} \qquad (10)$$

In this way, the final equation used to treat the data with non-linear least squares method, is as follows:

$$\sum \left(N_{ev} - S_C \frac{k_0 P T^{1/2}}{R(L-s)} ln\left(1 - \frac{10^{A-\frac{B}{C+T}}}{P}\right) \right)^2 \to min \qquad (11)$$

After the non-linear least squares is applied to treat the TG-data, the parameters of Antoine equation, describing dependence of vapor pressure on the temperature, are determined, as well as diffusion coefficient parameter. This procedure assumes that the molar mass of the substance in the vapor phase is known. The main advantage of this method is the possibility to determine vapor pressure and diffusion coefficient without calibration constant and reference substances. This eliminates the necessity to select the reference substance with reliable vapor pressure data, which are in deficit for low volatile compounds. The disadvantage of this method is that it is not applicable to $L = s$.

Very detailed and systematic study of process of evaporation of low volatile compounds in the presence of purge gas was carried out by Focke [63, 67]. In order to explain the observations of Gückel and Dollimore, where the dependence of the evaporation rate on the purge gas rate and low value of accommodation coefficient were detected, Focke developed the equation, describing the evaporation into the purge gas. His assumptions were the following:

• Constant physical properties. All physical properties, including the diffusion coefficient, D_{AB} are assumed concentration independent. The mole fraction of compound A anywhere in the purge gas is so low that the physical properties of the gas mixture are essentially identical to those of the pure purge gas.

- No chemical reaction occurs. The possibility of association of sample molecules in the gas phase is not considered.

- Gas solubility. It is assumed that the carrier gas is insoluble in the sample liquid/solid.

- Isothermal conditions. The thermogravimetric experiment is conducted under steady state isothermal conditions.

- Sample dimensions. The sample has well-defined dimensions with a characteristic length scale denoted by L. The size of the sample substance is assumed small compared to dimensions of the experimental cavity.

- Steady state conditions. In reality vaporization leads to a loss of sample volume and strictly speaking one is dealing with a moving boundary problem. It is assumed that the boundary regression is so slow that a pseudo-steady state assumption is valid.

- Ideal gases. All the vapors behave as ideal gases.

- Concentration of the sample substance A in the gas. The approaching purge gas does not contain compound A as an impurity. At the sample surface the concentration of compound A equals the equilibrium concentration at the prevailing temperature and pressure and is calculated from Raoult's law.

Based on these assumptions and mass balance of evaporating substance he derived the following equation for the analysis of TGA data:

$$p = \frac{RT}{k_c \cdot M} \frac{dm}{dt} \qquad (12)$$

k_c is mass transfer coefficient, defined by the geometry of measuring system, diffusion coefficient of the substance in the purge gas and the rate of purge stream. In case of diffusion into the stagnant gas the equation (12) transforms into:

$$p = \frac{RT}{MD} \frac{z}{S_C} \frac{dm}{dt} \qquad (13)$$

where z is diffusion path length, equal to the height of crucible, occupied by gas. In this way, comparison of the equation (13) with Langmuir equation (3) gives the expression for accommodation coefficient:

$$\alpha = \frac{D}{z}\sqrt{\frac{2\pi M}{RT}} \tag{14}$$

The value of α, calculated from Eq. (14), is in close agreement with that, obtained by Dollimore [64]. The method developed by Focke gives a good description of the evaporation process into the purge gas. It clearly shows that accommodation coefficient is not substance independent and the procedure proposed by Hawkins and Price [59, 32] is not applicable in general. Also, it explains that the method by Dollimore [64] for evaluation of vapor pressures is more preferable, since it requires the structural similarity of the investigated and the reference compounds. However, the method by Focke [63] requires data on diffusion coefficient of the substance in the purge gas, which are not readily available.

In the recent years Vecchio et al [65, 68, 69, 70] have used TGA with non-isothermal and isothermal conditions for measurements of vaporization enthalpies of different classes of organic compounds (substituted acetanilides, simetryn, ametryn, terbutryn, , and other pesticides). In all studies the vaporization coefficient α was considered to be constant. Careful validation of TGA procedure and uncertainties assessment has been performed using benzoic, succinic, and salicylic acids. The TGA data have been shown to be in agreement with the recommended values, provided that the study is performed very narrow temperature range (about 30 K) [65].

Experimental procedures reported for TGA have been developed mostly for studies of solid molecular compounds. What about liquid and ionic compounds? Are there any peculiarities in comparison to solid and molecular compounds to be expected?

Luo et al [85] were the first who applied TGA for the determination of the enthalpies of vaporization for ionic liquids. They applied isothermal conditions for determination of the vaporization rates. In contrast to non-isothermal TGA method, during each isothermal step the very low vaporization rates can be detected with sufficient uncertainty and reproducibility. In addition, the higher sensitivity at the isothermal step allows perform experiments at substantially lower temperatures where a possible decomposition of the IL sample is hardly expected. But it should be mentioned that the use of the isothermal steps during investigation consumes essentially more time in comparison the non-isothermal scanning mode.

Also Luo et al [85] were the first who applied TGA for ILs, their enthalpies of vaporization for the [$C_n$mim][NTf$_2$] series spread within (5-7) $kJ \cdot mol^{-1}$ around the data obtained by another established methods: Knudsen [72], transpiration [73], and TPD-MS [71, 74, 75]. Unfortunately, not much detail on the experimental procedure was reported in this work [85] in order to understand the reason for the data scatter. May be the very narrow temperature range (45 K) and only 4 isotherms have caused this scatter.

Jess and coworkers [76, 77] used non-isothermal TGA technique in order to study the long-term stability of different types of ILs. They have suggested an excellent analysis of the mass loss process occurring in the conditions of the TGA measurements, Using purge gases with different diffusion properties they were able to separate the vaporization and decomposition processes in the TGA run. They found that for [C_4mim][NTf_2] the rate of decomposition is very low up to 653 K. They derived enthalpy of vaporization of 108 $kJ \cdot mol^{-1}$ (at 600 K) for [C_4mim][NTf_2]. This value is in agreement with the results obtained by Knudsen [72] and mass-spectrometry [71, 74, 75]. However, this method is very sophisticated and needs a large amount of auxiliary data. It can be hardly possible to use as an easy and fast method to obtain enthalpies of vaporization of ILs.

Aschenbrenner et al [78] used isothermal TGA method to estimate the vapor pressure of some ILs: [C_4mim] [NTf_2], [C_2mim] [C_2SO_4], [C_6mim] [CF_3SO_3], and poly [(p-vinylbenzyl) - trimethylammonium hexafluorophosphate]. Glycerol was used as a reference compound in the determination. Their vapor pressure for [C_4mim][NTf_2] is three orders of magnitude higher than those from well established Knudsen technique [72]. For other ILs the vapor pressure was assessed to be near 1 to 2 Pa at 373 K. These values seem to be generously overestimated. That is why we are very reticent to use TGA for determination of the absolute vapor pressures without careful elaboration of the experimental procedure. In contrast, the TGA method is apparently very promising for the express, but reliable determination of vaporization enthalpy.

Summing up the available details on the TGA experiment, the isothermal mode is apparently more preferable for the reliable determination of vaporization enthalpies. However, a careful elaboration of the TGA procedure towards reproducible measurements for ILs is still required. In order to adjust the TGA experimental conditions for extremely low volatile compounds the screening of the limits for the following parameters is necessary:

1. amount of sample evaporated during each isothermal step;
2. flow rate of the purge gas;
3. temperature range for the determination of the vaporization rate.

Using optimized parameters it could be possible to adjust TGA for express determination of vaporization enthalpies in agreement with the other methods within 2-3 $kJ \cdot mol^{-1}$. In our opinion, the more precise determination of the enthalpy of vaporization lies outside the limits of the method of the fast evaporation from the open surface used in TGA.

It is reasonable to perform screening of the TGA parameters using some reference compounds with the well established pressures and vaporization enthalpies. As a matter of fact, only a few very low volatile compounds with the reliable data are available in the literature. In this work we have used hexadecane, di-butyl phthalate, methyl, and ethyl stearates for screening the experimental conditions.

3.2 Experimental section

3.2.1 Materials and chemicals

Samples of reference materials (hexadecane, di - butylphthalate, methyl stearate, and butyl stearate) were of commercial origin and they were distilled under reduced pressure prior to use and stored under dry nitrogen. The degree of purity was determined using a gas chromatograph (GC) equipped with a flame ionization detector. The carrier gas (nitrogen) flow was 7.2 $dm^3 \cdot h^{-1}$. A capillary column HP-5 (stationary phase crosslinked 5% PH ME silicone) was used with a column length of 30 m, an inside diameter of 0.32 mm, and a film thickness of 0.25 µm. The standard temperature program of the GC was T = 323 K for 180 s followed by heating to T = 523 K with the rate of 10 $K \cdot min^{-1}$. No impurities (greater than 0.05 mass per cent) could be detected in the samples used for the measurements.

Samples of [C$_2$mim][NTf$_2$], [C$_4$mim][NTf$_2$], [C$_6$mim][NTf$_2$], [C$_8$mim][NTf$_2$], [C$_{10}$mim][NTf$_2$] of commercial origin (IoLitec) were used. Samples of [C$_{12}$mim][NTf$_2$], [C$_{14}$mim][NTf$_2$], [C$_{16}$mim][NTf$_2$] samples were synthesized by Prof. P. Wasserscheid. Samples of ILs from [C$_2$mim][NTf$_2$] to [C$_{12}$mim][NTf$_2$] were transparent, colorless or slightly yellow liquids without odor. [C$_{14}$mim][NTf$_2$] and [C$_{16}$mim][NTf$_2$] were white crystals without odor.

Samples of [C$_2$mim][Br], [C$_4$mim][Br], [C$_2$mim][Cl], and [C$_4$mim][Cl] of commercial origin (Iolitec) were used. Samples of [C$_1$mim][Cl] and [C$_6$mim][Br] were synthesized by Dr. Tim Peppel and Prof. M. Köckerling (University of Rostock). Structures of the non-commercial samples were proved by C^{13} NMR. Prior to use IL samples were subjected to vacuum evaporation to remove possible traces of solvents and moisture. Samples were additionally purified in the preconditioning experiments.

3.2.2 TGA — Measurements of Vaporization Enthalpy

In this work we used Perkin Elmer Pyris 6 TGA, The calibration was performed with the use of extra pure (99.999%) alumel, nickel, and perkalloy by means of registration of Curie-points. The uncertainty of temperature calibration was less than 0.5 K. Mass calibration was performed with the use of standard weight of 100 µg. The purge gas -nitrogen was obtained by the vaporization of the liquid nitrogen from a Dewar vessel.

About 50-60 mg of a sample under study was placed into the platinum crucible with the vertical walls, diameter of 10 mm and the height 3 mm. In order to keep the evaporation area possibly constant the crucible was filled up to the edge. The filling station was organized in the glove box. Solid samples were solved in dry acetonitrile and charged to the crucible with a syringe. Then, the solvent was evaporated at the 100 °C for 2 hours. Prior to measurement of the vaporization enthalpy, a careful conditioning of the sample inside the TGA have been performed. During the conditioning procedure the sample was several times heated and cooled down in scanning mode in order to remove any traces of the solvent and the occluded water. The conditioning was repeated until the reproducible mass loss within two consequent runs was recorded.

The balance of the TGA instrument is highly sensitive but at the same time is subject for possible systematic errors due to the thermal drift or purge gas rate fluctuations. That is why it is vitally important to optimize measuring conditions for minimizing the interfering factors. The main parameters responsible for reliable vaporization enthalpy measurements using the TGA are discussed below.

3.2.2.1 Mass uptake

Mass uptake is the mass of the sample evaporated during each isotherm. Using in the TGA experiments very small mass losses could substantially reduce measuring times at lower temperatures. This is essentially important for the extremely low volatile ionic liquids. But, if the mass uptake is comparable with uncertainty of the balance, the reliable result is hardly expected. The lower level for mass uptake is determined by the longtime stability of the specific TGA setup (see Fig. 3.2). The uncertainty of the mass determination by the Perkin Elmer Pyris 6 is 0.02 µg. From a series of experiments we have concluded that the mass uptake during each isotherm should be at least more than 0.06 µg.

The upper limit of mass uptake is stipulated by an impact of the mass change inside the crucible on the surface area and the distance between vaporization surface and gas stream. In order to localize the maximal mass uptake during the isothermal step, the series of experiments were carried out. The enthalpies of vaporization for the reference compounds (hexadecane, di-butyl phthalate, methyl and ethyl stearates) were determined at the flow rate of nitrogen 60 mL·min⁻¹, the temperature interval of the enthalpy of vaporization determination was 60 K, the duration of each isotherm was at least 10 min. The mass loss during isothermal run was changed stepwise from 0.2 mg to 1.4 mg. The results are shown in Figure 3.3.

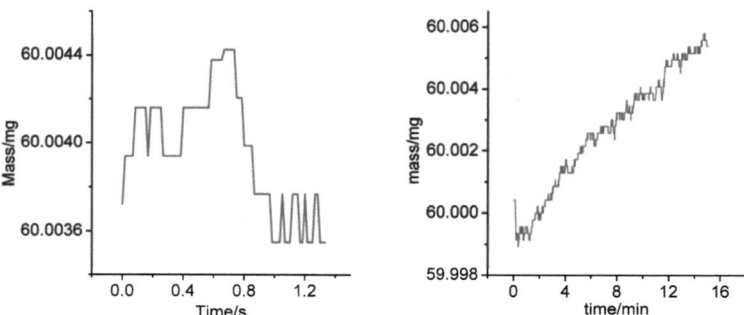

Figure 3.2: Digital noise and longtime stability for Perkin Elmer Pyris 6 TGA at T=303 K and N$_2$-flow is 140 *ml • min⁻¹*.

It is apparent from these figures that having the mass loss from 0.2 to 0.8 mg during the isotherm we were able to reproduce enthalpies of vaporization of the reference compounds within ±1 *kJ • mol⁻¹*. The mass uptakes larger than 1 mg lead to systematically higher vaporization enthalpy values for all reference compounds studied.

Conclusion: the optimal mass loss is 0.2 to 0.8 mg at each isothermal step.

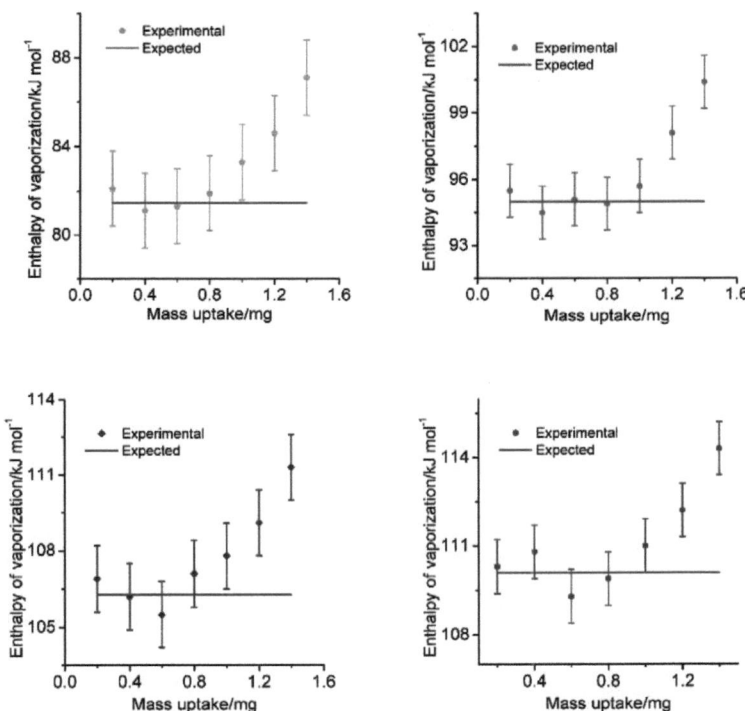

Figure 3.3: Enthalpies of vaporization of reference substances, obtained at different mass losses.
● Hexadecane ● di-Butyl Phthalate ● Methyl Stearate ◆ Ethyl Stearate

3.2.2.2 Duration of the isothermal step

Another important factor affecting the proper measurements of the mass uptake is a temperature stabilization time during the isothermal run. This factor is specific for each TGA setup. For the Perkin Elmer Pyris 6 TGA the temperature jump to the next level even with a slow heating (1 K · min^{-1}) has required at least 4 minutes for T-stabilization (see Fig. 3.4). It means that recording of the mass loss during this period should be excluded from the data treatment. This fact is especially important to consider at the high temperatures where due to high vaporization rates the experiment is usually reduced to about 10 minutes (time enough to reach up to 0.8 mg of the mass loss).

Conclusion: the duration of the experiment at each isotherm have to take into account the T-stabilization interval.

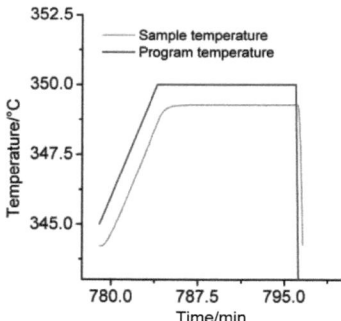

Figure 3.4: Temperature stabilization profile for Perkin Elmer Pyris 6 TGA

The optimal temperature range for the enthalpy of vaporization determination is specific for each compound dependent on its volatility. Thus, only general guidelines for minimal and maximal temperatures could be suggested. All TGA instruments show certain drift of the base line of the signal under isothermal conditions, thus vaporization rate should overcome this drift and the vaporization rate should be large enough to minimize the influence of the drift. The minimal vaporization rate related to unit surface was found to be $6.3 \cdot 10^{-8}\ kg \cdot s^{-1} \cdot m^{-2}$ for Pyris 6 TGA. Following, the minimal temperature of the experiment should correspond to this value.

The maximal possible temperature of the experiment is usually defined by the thermal stability of the sample. Good practice to determine a decomposition temperature is to perform dynamic TGA at several heating rates and to extrapolate the onset to the zero heating rate. An actual temperature interval has to be at least 60 degrees. The large temperature range is more preferable. As a rule, the experimental time to collect 0.06 mg mass loss and the long-term stability are responsible for the lowest temperature of TGA runs. To save time it is reasonable to shift the study to the elevated temperatures. However, shifting the study to higher temperatures has as a consequence the larger distance of extrapolations of obtained vaporization enthalpies from the average temperature T_{av} to the reference temperature 298.15 K. Another thing, which is indispensable to take into account, is the length of the interval.

Conclusion: the optimal temperature range should be at least 60 K and it limits depend on the volatility of the sample under study.

The main task of the purge gas is to sweep the evaporated sample out of the crucible. At low flow rates of the purge gas the saturation of the gaseous phase above the sample could aggravate the mass transport. In these conditions the mass loss occurs mainly due to the diffusion from the open surface into the gas phase and the Langmuir equation (3) is not valid for this case. At high rates the surface of the liquid sample is exposed to the turbulence of the gas flow. The temperature maintaining of the sample could also be affected.

The influence of the N_2-flow rates on the vaporization rate and the enthalpy of vaporization is shown in Fig. 3.5 and 3.6. It has turned out that the flow rates within 140 to 200 $ml \cdot min^{-1}$ provide optimal experimental conditions.

Conclusion: the gas flow rate of 140 $ml \cdot min^{-1}$ is optimal for Perkin Elmer Pyris 6 TGA.

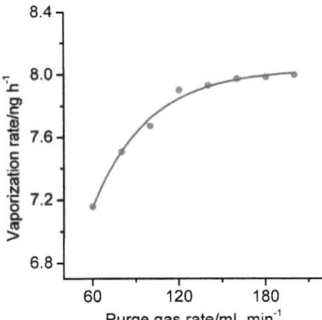

Figure 3.5: The influence of purge gas rate on the evaporation rate: di-butyl phthalate at 363 K.

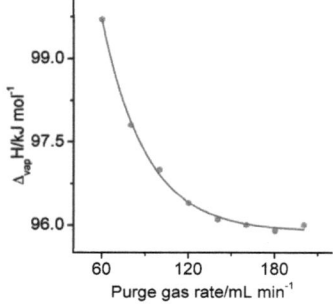

Figure 3.6: The influence of purge gas rate on the measured enthalpy of vaporization of di-butyl phthalate.

Optimization of the experimental parameters is required for each commercially available type of TGA. When using the Perkin Elmer Pyris 6 TGA the following experimental conditions have to be kept for the reliable determination of vaporization enthalpies of molecular compounds:
- mass uptake at each temperature – 0.2-0.8 mg (for molecular compounds)
- duration of isothermal steps – at least 10 min
- temperature range at least 60K
- purge gas – 140 to 200 $mL \cdot min^{-1}$

3.3 Results

3.3.1 Test-measurements of vaporization enthalpies of the reference Molecular Liquids

Having established the optimal conditions for the TGA experiment, the set of reference compounds: hexadecane, di-butyl phthalate, methyl, and ethyl stearates has been studied. These molecular compounds, which are heavy volatile, have well established vaporization enthalpies. It is possible to test the selected experimental conditions with these molecular compounds in order to extend these conditions for experiments with ionic liquids, where reliable experimental data are still absent [17]. For each compound at least 5 to 7 TGA-runs in the selected T-range have been performed. The first (and sometimes the second) run have not been taken into account because of conditioning of the sample as described above. The resultative runs were treated together. The mass loss values measured in the different runs were scaled due to decreasing of the absolute value of the sample from run to run.

3.3.2 Data treatment of the TGA-measurements of molecular liquids

Enthalpies of vaporization derived using Eq. (7) are referred to the mean temperature of the studied range. These values have to be adjusted to the reference temperature of 298.15 K using the following equation:

$$\Delta_{vap}H_{298} = \Delta_{vap}H_T - \Delta_l^g c_p \cdot (T - 298.15) \qquad (15)$$

where, $\Delta_l^g c_p$ is the difference of heat capacities of liquid and gas phases.

Values of $\Delta_l^g c_p$ have been derived from the isobaric molar heat capacities of the liquid compounds, $c_{p,l}$ ($J \cdot mol^{-1} \cdot K^{-1}$), calculated according to the group additivity method procedure developed by Chickos [79]:

$$\Delta_l^g c_p = 10.58 + 0.26 \cdot c_{p,l} \qquad (16)$$

The groups and their contributions, as well as heat capacities of liquid phase and heat capacities differences are listed in Table 3.3.

Table 3.3: Estimation of heat capacity differences using group additivity of Chickos [80].

Substance	Groups	Group parameters	$c_{p,l}^0$	$\Delta_l^g c_p$	$\Delta_l^g c_p^{exp\ a}$
Hexadecane	$CH_3 - 2$	34.9	516.4	144.8	134.2±12.7
	$CH_2 - 14$	31.9			
di-Butyl phthalate	$CH_3 - 2$	34.9	505.4	142.0	148.1±14.2
	$CH_2 - 6$	31.9			
	$C_BH - 4$	21.8			
	$C_BR - 2$	15.3			
	$C-COO - 2$	63.2			
Methyl stearate	$CH_3 - 2$	34.9	643.4	177.9	158.9±16.9
	$CH_2 - 16$	31.9			
	$C-COO - 1$	63.2			
Ethyl stearate	$CH_3 - 2$	34.9	675.3	186.2	179.8±18.4
	$CH_2 - 17$	31.9			
	$C-COO - 1$	63.2			

[a] - Data from Table 3.5. Heat capacities are in $J \cdot mol^{-1} \cdot K^{-1}$

Comparison of the enthalpies of vaporization obtained from TGA with those from other techniques is given in Table 3.5. The errors given in this table are twice standard deviations of the mean values.

Table 3.5: Reference molecular compounds. Comparison of the enthalpies of vaporization (at 298.15 K) measured using TGA and other techniques.

Substance	Temperature range, °C	B^a	C^a	$\Delta_l^g c_{p,exp}^0$ $J \cdot mol^{-1} \cdot K^{-1}$	$\Delta_{vap} H_{298}^{exp}$ $kJ \cdot mol^{-1}$	$\Delta_{vap} H_{298}^{lit}$ $kJ \cdot mol^{-1}$	Reference method
Hexadecane	50-110	14567.2	-16.141	134.2±12.7	81.1±1.7	80.6[81]	ebulliometry
di-Butyl phthalate	90-150	16869.9	-17.813	148.1±14.2	95.0±1.2	94.0[82]	transpiration
Methyl stearate	50-110	18472.0	-19.112	158.9±16.9	106.2±1.3	104.0[82]	transpiration
Ethyl stearate	90-150	19666.5	-21.626	179.8±18.4	109.9±0.9	110.8[82]	transpiration

^{a} – see Eq. (18)

As can be seen from the Table 3.5 in the selected experimental conditions we have been able to reproduce the recommended literature values for very low volatile compounds within =2.0 $kJ \cdot mol^{-1}$.

However, in order to measure to the absolute values of vapor pressure using TGA an extended study of the vaporization coefficient α is required.

In order to demonstrate the importance of the further careful studies, the absolute vapor pressures of hexadecane measured using the transpiration technique are compared with the results from TGA (Fig. 3.7). As a matter of fact the mass uptake directly measured by TGA is only proportional to the values of $\ln p$. According to the Langmuir equation (3) the vaporization coefficient is responsible for the adjustment of the results to the absolute vapor pressure. In the flow of the purge gas α can be considered as a constant. However, the available in the literature information on α the is contradicting. Moreover, until now is nothing known about α in case of ILs. For example: is it the same within the homologous series or is it different for the molecular and the ionic compounds. The nature of this coefficient will be carefully studied using comparison of the TGA data with the absolute vapor pressures obtained e.g. static or transpiration method, which are well established in our laboratory. This work is still in progress.

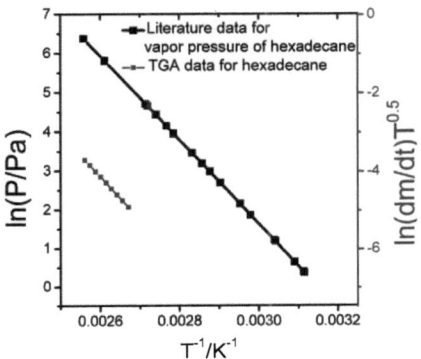

Figure 3.7: Comparison of the results from transpiration and TGA for hexadecane.

3.3.3 Test-measurements of vaporization enthalpies of Ionic Liquids

With the reliable data on vaporization enthalpies for the reference compounds (molecular liquids) we have tested influence of all important experimental conditions (T-range, mass uptake, flow-gas rate, evaporation surface, initial mass of sample, conditioning of the sample etc.). These studies have allowed to localize the boundaries for variation of experimental parameters and to recommend the selected conditions for the reliable determination of the vaporization enthalpy for the heavy volatile molecular compounds. As can be seen from the Table 3.5, it is possible to measure $\Delta_{vap}H$ (298 K) of the reference compounds within ± 1 to 2 $kJ \cdot mol^{-1}$ by using TGA.

Having established the optimal conditions for the molecular compounds, we have applied them for investigation of the ionic liquids [$C_n$mim][NTf$_2$], where numerous vaporization enthalpies are already available [17]. Vaporization enthalpies of ILs have been measured using the TGA unit - Perkin Elmer Pyris 6 TGA. Isothermal gravimetric analysis curves were measured in the temperature range 480-570 K in nitrogen at a flow rate of 140 $mL \cdot min^{-1}$. Platinum plane crucible (diameter 10^{-2} m) with about 70 mg of the sample was used. A heating ramp of 10 $K \cdot min^{-1}$ was used, followed by a 4 h static hold period at 150°C, allowing for the slow removal of volatile impurities such as 1-methylimidazole and traces of water prior to a stepwise isothermal run. For our experiments, we used the same platinum crucible for each sample, in order to maintain a uniform cross-sectional area. Already in preliminary measurements it have been shown that due to very low volatility of ILs the mass uptake of only 0.06-0.1 mg is possible to obtain at each temperature.

However such small mass uptake still fits the recommended experimental conditions (see Chapter 3.2.2).

The absence of decomposition during the vaporization process was verified by ATR IR spectroscopy of the initial sample, sample condensed on the apparatus lid and sample left in the crucible after enthalpy of vaporization determination No evidence for thermal degradation has been observed, consistent with results from our earlier measurements [38]. The IR spectra of [C_{12}mim][NTf_2] is shown in Figure (3.8) as an example. The differences in obtained spectra are connected to small amount of the sample condensed on the cover, but not to any possible decomposition.

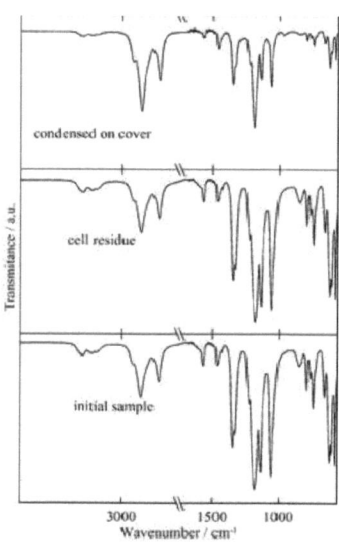

Figure 3.8: The spectra of [C_{12}mim][NTf_2] used in TGA investigations [83].

3.3.4 Data treatment of the TGA-measurements of ionic liquids

Enthalpies of vaporization of ionic liquids derived using equation (7) from vaporization rates measured from TGA are given in Table 3.7. These enthalpies of vaporization are also referred to the mean temperature of the studied range and they have to be adjusted to the reference temperature of 298.15 K using equation (15). Adjustment of the vaporization enthalpy to the reference temperature is usually performed under the assumption of the constant difference between heat capacity liquid and gaseous phases. In contrast to the molecular compounds, values of for ionic liquids are ill defined. Because of the total lack of the information, the arbitrary values of of 94 $J \cdot mol^{-1} \cdot K^{-1}$, 100 $J \; mol^{-1} \; K^{-1}$ and 105 $J \; mol^{-1} \; K^{-1}$ [72] are commonly used for ILs regardless on their structures. In this work we suggest to apply the Clarke and Glew equation to obtain the heat capacity difference from our experimental TG–data. Clarke and Glew [84] suggested to approximate the vapor pressure temperature dependence using the three adjustable parameters with the physical meaning as enthalpy $\Delta_{vap} H^0_{298}$, and entropy $\Delta_{vap} S^0_{298}$ of vaporization, as well as the difference between heat capacity of liquid and gaseous phases $\Delta_l^g c_p$:

$$\ln p = \frac{\Delta_{vap} S^0_{298}}{R} - \frac{\Delta_{vap} H^0_{298}}{RT} - \frac{\Delta_l^g c_p}{R}\left(\ln \frac{T}{298.15} - 1 + \frac{298.15}{T}\right) \quad (17)$$

Combination of equations (3) and (17) gives the following relationship:

$$n\left(\frac{dm}{dt}\sqrt{T}\right) = A - \frac{B}{T} - C \cdot \ln\left(\frac{T}{298.15}\right) \quad (18)$$

where $\Delta_l^g c_p = C \cdot R$ and $\Delta_{vap} H^0_{298} = B \cdot R - \Delta_l^g c_p^0 \cdot 298.15$.

We have tested this procedure for the reference molecular liquids (see Table 3 and 4) successfully. The results of the determination of $\Delta_{vap} H$ at the average and reference temperature are listed in Table 3.7.

Table 3.7: The enthalpies of vaporization for 1-alkyl-3-methylimidazolium bis – (trifluoromethanesulfonyl) – imides by TGA method.

n in [C$_n$mim][NTf$_2$]	$\Delta_{vap} H$ (542 K) $kJ \cdot mol^{-1}$	$-\Delta_l^g c_p$ $J \cdot mol^{-1} \cdot K^{-1}$	$\Delta_{vap} H$ (298.15 K) $kJ \cdot mol^{-1}$	$\Delta_{vap} H$ (380 K) $kJ \cdot mol^{-1}$ TGA	$\Delta_{vap} H$ [a] (380 K) $kJ \cdot mol^{-1}$ QCM
2	109.5±2.0	45.8±5.7	120.8±2.4	116.9±2.2	118.5±1.0
4	113.4±2.0	65.7±7.3	129.5±2.7	124.1±2.3	124.3±1.0
6	117.9±2.0	94.1±9.5	140.9±3.1	133.2±2.5	131.9±1.0
8	122.6±2.0	125.0±11.3	153.2±3.4	142.9±2.7	137.5±1.0
10	122.6±2.0	142.6±15.7	157.1±4.0	145.8±3.2	144.3±1.0
12	124.7±2.0	159.6±18.1	163.6±4.7	150.6± 3.5	151.1±1.0
14	132.6±2.0	169.1±20.4	173.9±5.4	160.1±3.9	158.4±1.0
16	136.3±2.0	175.6±21.3	179.2±6.0	164.8±4.6	164.4±1.0

[a]- Data from Knudsen QCM method [83].

As a rule for each sample under study a large amount of experimental points are obtained. The standard deviation of the experimental vaporization enthalpy was usually within 0.8 to 1.8 $kJ \cdot mol^{-1}$. However, taking into account that TGA method is not a method developed for the enthalpy of vaporization determination, the standard deviation represents only the reproducibility of the TGA experiment. It is difficult to assess the systematic error inherent for TGA. Therefore, in case when the enthalpy of vaporization is good reproducible within ±1 $kJ \cdot mol^{-1}$ we are reticent to ascribe this very small uncertainty to the measured value, because it means generous overestimation of the reliability the result. Nevertheless, from our experiences with measurements of the reference molecular compounds, we have observed the agreement with the recommended data on $\Delta_{vap}H$ within ± (1 to 2) $kJ \cdot mol^{-1}$. Therefore, for the enthalpies of vaporization for ILs measured by TGA the uncertainty of the result was estimated as doubled standard deviation. When the estimated in such way uncertainty was lower than 2 $kJ \cdot mol^{-1}$, it was taken to be equal 2 $kJ \cdot mol$.

In uncertainties of enthalpies of vaporization adjusted to 298.15 K the uncertainty of $\Delta_{vap}H_T$ at mean temperature and the uncertainty of the heat capacity difference between liquid and gaseous state $\Delta_l^g c_p$ were included according to the following equation:

$$\Delta\Delta_{vap}H_{298} = \sqrt{\Delta\Delta_{vap}H_T^2 + \left(\Delta\Delta_l^g c_p \left(T - 298.15\right)\right)^2}$$

3.4 Discussion

ILs are claimed to have negligible vapor pressures and most of the chemical and technical applications of ILs are based on this remarkable property. A reasonable accuracy of the TGA method for measurements of vaporization enthalpies of imidazolium based ILs have been demonstrated just recently [77, 85] and it has prompted the current study. At the same time the recent paper by Styring et al. [78] have demonstrated quite clearly complications which are arising, when the vapor pressure data are derived from the TGA data. Their vapor pressure for [C$_4$mim][NTf$_2$] of 1.2 Pa at 120°C is 12000 times higher compared to the result expected at this temperature from the Knudsen measurements [72]. Taking into account these advantages and complications we decided first to develop the TGA method to measure enthalpies of vaporization of the very low volatile liquids (molecular and ionic) accurately.

3.4.1 TGA studies of 1-alkyl-3-methylimidazolium bis (trifluoromethane sulfonyl) imides

Already previous investigations of ILs have shown that 1-alkyl-3-methylimidazolium bis (trifluoromethanesulfonyl) imides $[C_n\text{mim}][NTf_2]$ were remarkably stable to be vaporized without decomposition at rather high temperatures of 300 to 573 K [85, 74, 86, 17, 72, 75]. However the available in the literature data on vaporization enthalpies for this homological series are apparently in disarray (see Fig. 3.9 and Table 7). There are different tendencies in the enthalpies of vaporization for this series measured in the different works. For example, an unusual for homologous series non-linear dependence of the enthalpy of vaporization on the chain length of the in the cation was observed for ethyl-, butyl-, hexyl- and octyl- derivatives determined by the Knudsen method [72] and by TPD MS [74]. Also from the results of Knudsen method [72] the vapor pressure for $[C_6\text{mim}][NTf_2]$ is higher than that for $[C_4\text{mim}][NTf_2]$. At the same time, the direct calorimetrically measured values of vaporization enthalpies [17] showed the expected liner dependence of the enthalpy of vaporization. However, the contribution for CH_2 group to the enthalpy of vaporization from the calorimetric study was surprisingly high (8.9±0.6 $kJ \cdot mol^{-1}$) in comparison to the same contribution to the enthalpy of vaporization in n-alkanes (4.95 ± 0.10 $kJ \cdot mol^{-1}$) [87]. Moreover, for the most frequently studied by different methods sample of $[C_2\text{mim}][NTf_2]$ the spread of the available data (Fig. 3.10) is too large for any reasonable interpretation of this results.

Therefore, taking into account the very good thermal stability of the $[C_n\text{mim}][NTf_2]$ series and plethora of available experimental data for these ILs, they seems to be of the first choice for TGA study in this work.

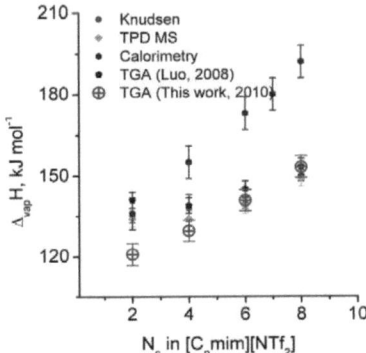

Figure 3.9: Experimental data for homologous series [C$_n$mim][NTf$_2$] available in the literature.

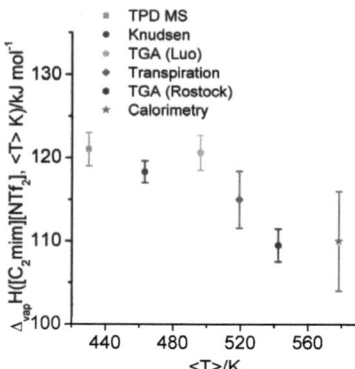

Figure 3.10: Enthalpy of vaporization of [C$_2$mim][NTf$_2$] from different techniques.

The general formula for the vaporization enthalpy calculations of ILs at 298 K is:

$$\Delta_{vap}H_{298} = \sum_i n_i \cdot H_i + \sum_j n_j \cdot H_j \qquad (19)$$

where H_i is a contribution of the z-th element; n_i is the number of elements of the z-th type in the ionic liquid; H_j is a contribution of the j-th structural correction; rt_j is the number of structural corrections of the j-th type in the ionic liquid. In the simple case for the [C$_2$mim][Et$_2$SO$_4$] (empirical formula C$_8$H$_{16}$N$_2$SO$_4$) the calculation formula is:

$\Delta_{vap}H([C_m mim][Et_2SO4]) = 8 \cdot He + 2 \cdot H_N + 4 \cdot NO + 1 \cdot NS$

Table 3.9: Vaporization enthalpies $\Delta_{vap}H$ of ILs at 298 K - experiment and prediction.

Anion	Cation	Method	$\Delta_{vap}H_T^{exp}/T$, kJ·mol^{-1} / K	$\Delta_{vap}H_{298.15}^{exp}$, kJ·mol^{-1}	$\Delta_{vap}H_{298}^{pred\,a}$, kJ·mol^{-1}	exp − pred, kJ·mol^{-1}
[C$_2$mim]	[Et$_2$SO$_4$]	TPD MS [71]	145±4.0/500	164±4.0	163.8	0.2
[C$_2$mim]	[NTf$_2$]	Knudsen [72]	118.8 ±1.3/463	135.3 ±1.3	132.9	2.4
		Transpiration [73]	115.0±3.4 /519	136.7±3.4		4.1
		TPD MS [71]	121.0 ±2.0/430	134.0 ±2.0		1.1
		Calorimetry [88]	110±6 /578	136±6		3.1
		TGA [85]	120.6 ±2.1/496	141.0 ±2.1		8.1
[C$_4$mim]	[NTf$_2$]	Knudsen [13]	120±5 /487	138 ±5	137.9	0.1
		Knudsen [72]	118.3 ±1.7/478	136.2 ±1.7		-1.7

		TPD MS [71]	120.0 ±3.0/440	134.0 ±3.0		-3.9
		Calorimetry [88]	128±6 /578	155±6		17.1
		TGA [85]	118.5 ±0.4/496	139.0 ±0.4		1.1
[C_6mim]	[NTf_2]	Knudsen [72]	123.4 ±0.8/462	139.8 ±0.8	142.9	-2.9
		TPD MS [71]	124 ±2/445	139 ±2		-4.1
		Calorimetry [88]	145±6 /578	173±6		30.1
		TGA [85]	124.1 ±0.7/503	145.1 ±0.7		2.2
[C_7mim]	[NTf_2]	Calorimetry [88]	151 ±6 /578	180 ±6	145.4	34.6
[C_8mim]	[NTf_2]	Knudsen [72]	132.3 ±0.8/475	150.0 ±0.8	147.9	2.1
		TPD MS [71]	134 ±2 /450	149 ±2		1.1
		Calorimetry [88]	163±6 /578	192±6		44.1
		TGA [85]	132.3 ±0.5/503	153.3		5.4
[C_{10}mim]	[NTf_2]	TGA [85]	134.0 ±2.3/510	153.0 ±2.3	152.	0.1
[C_2mim]	[beti]	TGA [85]	115.3 ±1.8/503	136.3 ±1.8	192.7	-65.4
[C_4mim]	[beti]	TGA [85]	114.4 ±0.4/503	135.4 ±0.4	197.7	-62.3
[C_6mim]	[beti]	TGA [85]	118.4 ±1.6/503	139.4 ±1.6	202.7	-63.3
[C_8mim]	[beti]	TGA [85]	125.0 ±0.8 499	145.0 ±0.8	207.7	-62.7
[C_{10}mim]	[beti]	TGA [85]	127.4 ±3.7/510	148.4 ±3.7	212.7	-64.3
[C_8mim]	[OTf]	TPD MS [71]	133.0 ±3.0/498	151.0 ±3.0	125.7	25.3
[C_8mim]	[BF_4]	TPD MS [71]	141.0 ±3.0/520	162.0 ±3.0	160.4	1.6
[C_8mim]	[PF_6]	TPD MS [71]	147.0 ±4.0/530	169.0 ±4.0	168.9	0.1
[C_4mim]	[$N(CN)_2$]	Transpiration [73]	139.8 ±1.1/464	157.2 ±1.1	156.5	0.7
[C_8mim]	[$N(CN)_2$]	TPD MS [71]		162.0 ±4.0	166.5	-4.5
[$Pyrr_{1,4}$]	[$N(CN)_2$]	Transpiration [89]		160.0 ±2.0	159.8	0.2
[$Pyrr_{1,4}$]	[$PF_3(CF_3)_2$]	TPD MS[90]		152.0 ±4.0	152.3	-0.3
[C_2mim]	[NO_3]	Transpiration [91]		168.4	164.7	4.6
[C_4mim]	[NO_3]	Transpiration [91]		167.6	169.7	-0.8

[a] - evaluated using equation (19)

Enthalpies of vaporization of $\Delta_{vap}H$ for [$C_n$mim][NTf_2] series measured using TGA are given in Table 3.7. Most of these values were obtained at the average temperature 542 K. Any correlations at T = 542 K are preferable, because the enthalpies of vaporization are less affected by temperature adjustment with the arbitrary heat capacity difference values. In Fig. 3.11 the enthalpies of vaporization for [$C_n$mim][NTf_2] at 542 K are correlated with the chain length of the cation. Except for [C_{12}mim][NTf_2], the measured data are fitting well the expected linear correlation. However, the deviation of the data for [C_{12}mim][NTf_2] is not significant taking into account the boundaries of the experimental uncertainties of other members of this homologous series.

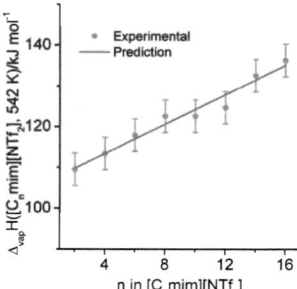

Figure 3.11: Dependence of the enthalpy of vaporization for 1-alkyl-3-methylimidazolium bis (trifluoromethanesulfonyl) imides at 542 K on the length of the cation alkyl chain.

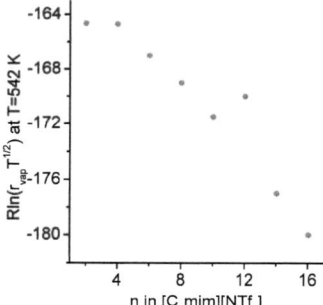

Figure 3.12: The dependence of the value analogue of vapor pressure for 1-alkyl-3-methylimidazolium bis (trifluoromethanesulfonyl) imides from the length of the alkyl chain

As it has been mentioned above, the absolute vapor pressures for [$C_n$mim][NTf$_2$] was not the goal of this study. But it is interesting to compare the vaporization rates for this series, because these parameters are actually "pressure-analogue" values *(P*)*. In Fig. 3.12 these *P**-values at 542 K are correlated with the chain length in [$C_n$mim][NTf$_2$] series. The vapor pressure analogue values *P** for [C_2mim][NTf$_2$] and [C_4mim][NTf$_2$] at 542 K are approximately equal (in agreement with the results for these two ILs quartz crystal microbalance method [83]). The P^* values for ILs with the longer chain show more or less linear behavior but again with the exception for the data point [C_{12}mim][NTf$_2$]. This outlier demonstrates a crucial sensitivity of the TGA conditions towards the absolute vapor pressure measurements. If for vaporization enthalpy measurements the data point for [C_{12}mim][NTf$_2$] was still in insignificant discrepancy (see Fig. 3.11), the deviation for the P^* - values seems to be already dramatic (see Fig. 3.12). For this reason the discussion of our results is focused on vaporization enthalpies of ionic liquids.

Experimental results from TGA measurements for [$C_n$mim][NTf$_2$] have been treated with equation (18) and the resulting enthalpies of vaporization adjusted to the reference temperature 298.15 K are given in Table 3.7 and in Fig. 3.13. For comparison, experimental $\Delta_{vap}H^o_{298}$ data for some molecular liquids (in homologous series) for alkyl benzenes *($C_nH_{2n+1}C_6H_5$)* [92], alkyl nitriles *($C_nH_{2n+1}CN$)* [93] and alcohols *($C_nH_{2n+1}OH$)* [94] are also shown in Fig. 3.13. It is apparent that slopes for the molecular and ionic liquids are quite similar and it means that contribution in the enthalpy of vaporization for the extension with CH_2 group (see Table 3.11) in ionic liquids is identical to those in molecular liquids. Comparison of the CH_2 increments to the enthalpy of vaporization of ILs and molecular liquids is given in Table 3.11. The increment for CH_2 group for molecular liquids lies between 4.44 to 5.03 kJ·mol^{-1}. The CH_2 increment for the enthalpy of vaporization of [$C_n$mim][NTf$_2$] series 4.2±0.4 kJ·mol^{-1} is somewhat lower but it is hardly ever specific within the experimental uncertainties. According to the X-ray studies of ILs liquid phase structure [95, 96] the ionic part and alkyl chain are separated, thus the van der Waals interaction and Coulomb interaction are not distorted by each other. Therefore, the CH_2 increment in ILs should be close to that in the molecule where a large functional group attached to the n-alkyl chain (e. g like alkyl benzenes). This theoretical finding is now proved by our experimental measurements (Table 3.11 and Fig. 3.13).

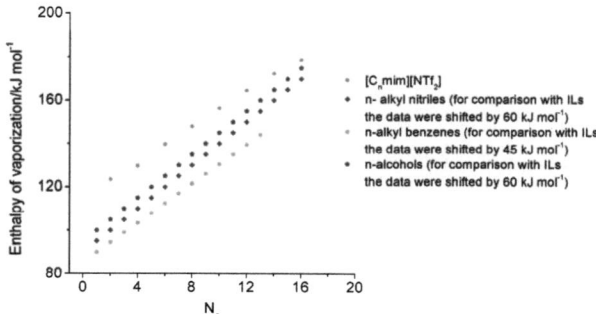

Figure 3.13: The enthalpies of vaporization for 1-alkyl-3-methylimidazolium bis - (trifluoromethanesulfonyl) - imides, alkyl benzenes, alkyl nitriles and alcohols at 298.15 K.

Table 3.11: The values of the CH2 group increment into the enthalpy of vaporization for the different classes of compounds.

Type of molecules	Increment value of $\Delta_{vap}H^0_{298}$, $kJ \cdot mol^{-1}$
$[C_n\text{mim}][NTf_2]$	4.2±0.4
$C_nH_{2n+1}CN$	4.44±0.12[93]
$C_nH_{2n+1}OH$	4.75±0.08[94]
$C_nH_{2n+1}C_6H_5$	4.49±0.06[92]
C_nH_{2n+2}	4.953[87]
$C_nH_{2n+2}CH=CH_2$	4.972[87]
$C_nH_{2n+1}SH$	4.760[87]
$C_nH_{2n+1}Cl$	4.854[87]
$C_nH_{2n+1}Br$	4.803[87]
$C_nH_{2n+2}CH=CH-CH_3$	5.029[87]

3.4.2 Validation of TGA vaporization enthalpies 1-alkyl-3-methylimidazolium bis (trifluorornethanesulfonyl) imides with Langmuir Quartz Microbalance Method

Rebelo *at al* [17] collected the available data for vaporization enthalpies of [$C_n$mim][NTf$_2$] series. Spread of these data is shown in Fig. 3.9. A new experimental technique Quartz Microbalance Method (QCM) has been just recently established at University of Rostock (Dr. Dz. H. Zaitsau, Prof. S.P. Verevkin) and this technique have been purpose-built for studies of ILs. In short: evaporation of an IL sample into vacuum is studied at several temperatures in an automatic Knudsen apparatus. In contrast to the traditional Knudsen technique the evaporation occurs from the open surface of the Knudsen cell and the Langmuir equation (3) is valid for such experimental conditions. In order to measure the mass loss the quartz crystal microbalance (QCM) is placed above the cell. The linear dependence between the resonance frequency change and deposited mass has been established and used for measurements of vaporization enthalpies.

New experimental data measured with this technique are collected in Tables 3.7 and 3.13. For comparison with these results we adjusted (using estimated according to Eq. (16) enthalpies of vaporization measured by TGA to 380 K (see Table 3.7), where he most reliable data for the [$C_n$mim][NTf$_2$] series were measured using Langmuir-QCM method recently. As can be seen from Table 3.7 the results from TGA and QCM are in very good agreement taking into account the combined experimental uncertainties. Only agreement for [C_8mim][NTf$_2$] is slightly worth in comparison to another representatives from this series. In our opinion, comparison of the data derived from different methods is always favorable at the elevated temperature related to the average temperature of experiment T_{av}. In this case the compared values are less affected by extrapolation to the desired temperature. Extrapolation of the measured vaporization enthalpies to the reference temperature is heavy aggravated with the ill defined values as discussed above. In spite of this fact, the $\Delta_{vap}H^0_{298}$ data are highly desired for development of MD simulations and first principles calculations. In this work we decided to extrapolate the TGA and QCM measured vaporization enthalpies to the reference temperature 298.15 K twofold: using the traditional way with the $\Delta_l^g c_p = 100\ J \cdot K^{-1} \cdot mol$ in the Eq. (15) and the second way is extrapolation using the values obtained in this work by data treatment with help of Eq. (18). Results are given in Table 3.13.

Table 3.13: Comparison of the data from TGA and QCM.

n in [C$_n$mim]	TGA		QCM		$\Delta_{vap}H^0_{298}$, kJ·mol^{-1} ($\Delta_l^g c_p = -100$, J·K^{-1}·mol^{-1})		$-\Delta_l^g c_p$	$\Delta_{vap}H^0_{298}$, kJ·mol^{-1}	
	T_m, K	$\overline{\Delta_{vap}H}$	T_m, K	$\overline{\Delta_{vap}H}$	TGA	QCM		TGA	QCM
1	2	3	4	5	6	7	8	9	10
2	532.1	110.6 ±2.0	378	118.6 ±1.0	134.0±2.0	120.8±2.4	45.8 ±5.7	121.3±3.3	122.3±1.5
4	542.4	113.5 ±2.0	378	124.4 ±1.0	137.9±2.0	129.5±2.7	65.7 ±7.3	129.5±3.8	129.6±1.6
6	542.4	118.0 ±2.0	383	131.6 ±1.0	142.4±2.0	140.9 ±3.1	94.1 ±9.5	141.0±4.3	139.6±1.8
8	542.5	122.6 ±2.0	387	136.8 ±1.0	147.0±2.0	153.2 ±3.4	125.0 ±11.3	153.1±4.8	147.9±2.0
10	522.4	125.2 ±2.0	394.6	142.5 ±1.0	147.6±2.0	157.1 ±4.0	142.6 ±15.7	157.2±5.5	156.3±2.5
12	532.1	126.0 ±2.0	408.4	147.0 ±1.0	149.4±2.0	163.6 ±4.7	159.6 ±18.1	163.3±6.2	164.6±3.0
14	542.4	132.7 ±2.0	416.2	152.5 ±1.0	157.1±2.0	173.9 ±5.4	169.1 ±20.4	174.0±7.0	172.5±3.4
16	542.5	136.4 ±2.0	424.8	156.3 ±1.0	160.8±2.0	179.2 ±6.0	175.6±21.3	179.3±7.2	178.5±3.7

As it can be seen from Table 3.13, with the constant heat capacity difference ($\Delta_l^g c_p = -100\ J\cdot K^{-1}\cdot mol^{-1}$), the disagreement between QCM and TGA results for all ILs sometimes exceed 10 $kJ\cdot mol^{-1}$ (see Table 3.13, column 6 and 7). However, the agreement of both data sets at 542 K and at 380 K has been just verified in Table 3.7. Following the large disagreement at 298.15 K between enthalpies of vaporization derived from TGA and QCM is solely due to the fact that for ILs under study is definitely not constant for the ILs with different chain length. In contrast, the values for [C$_n$mim][NTf$_2$] derived in this work using Eq. (18) (Table 3.13, column 8) monotonically rise with the chain length. It is also apparent from the $\Delta_{vap}H^0_{298}$ data presented in columns 9 and 10 (Table 3.13) that agreement between enthalpies of vaporization derived from TGA and QCM is in excellent agreement. Thus the values of for [C$_n$mim][NTf$_2$] (Table 3.13, column 8) could be recommended for re-calculation of the available data for this homologues series.

3.4.3 TGA studies of l-alkyl-3-methylimidazolium halides

Having established the express experimental TGA procedure, the most common ILs [$C_n$mim][Cl] and [$C_n$mim][Br] have been studied using this method. Results are given in Table 3.15.

It is well established, that thermal stability of imidazolium halides is substantially lower. The expected decomposition temperature is of 546 K according to the literature [97]. Thus, our investigation was carried out at significantly lower temperatures in comparison to [$C_n$mim][NTf$_2$] in order to exclude possible decomposition of the samples.

Enthalpies of vaporization of l-alkyl-3-methylimidazolium chlorides measured by TGA (column 5, Table 3.15) are in a good agreement with the results obtained from DSC studies (column 6, Table 3.15). At the same time, enthalpies of vaporization of l-alkyl-3-methylimidazolium bromides measured by TGA (column 5, Table 3.15) are systematically lower in comparison to results obtained from DSC and QCM studies. However from the IR-ATR spectra for the [$C_n$mim][Br] was apparent the decomposition of the sample during the TGA experiment. The typical spectra for [C_1mim][Br] is shown in Figure 3.14.

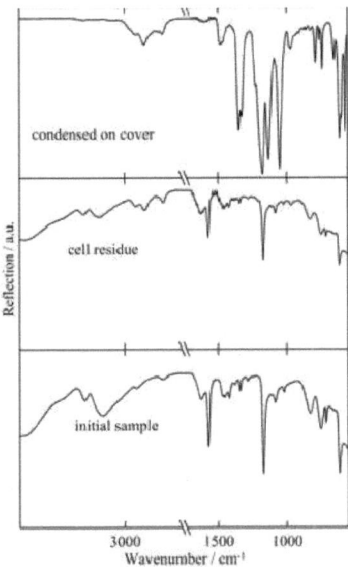

Figure 3.14: The IR-ATR spectra for [C_1mim][Br] used in enthalpy of vaporization determination by TGA method

Table 3.15: The enthalpies of vaporization for 1-alkyl-3-methylimidazolium halides by TGA method.

Sample	T_{av}, K	$\Delta_{vap}H^0_{T_{av}}$, $kJ \cdot mol^{-1}$	$-\Delta_l^g c_p$, $J \cdot mol^{-1} \cdot K^{-1}$	$\Delta_{vap}H^0_{298}$, $kJ \cdot mol^{-1}$ TGA	$\Delta_{vap}H^0_{298}$, $kJ \cdot mol^{-1}$ DSC/QCM
1	2	3	4	5	6
[C$_2$mim][Cl]	432.4	136.5±1.8	92±10	148.8±2.2	140.8±5.0[b]
[C$_4$mim][Cl]	432.4	137.2±0.9	122±16	153.6±2.3	152.4±3.0[b]
[C$_6$mim][Cl]	432.4	147.5±1.3	128±16	164.7±2.5	160.0±2.8[b]
[C$_1$mim][Br]	442.1	118.5±1.2	68±8	128.3±1.7[a]	141.4±5.0[b]
[C$_2$mim][Br]	432.4	117.1±1.8	96±10	129.9±2.2[a]	145.9±5.0[b]
[C$_4$mim][Br]	432.4	118.1±1.8	111±13	133.0±2.5[a]	156.8±2.4[b] 159.0±3.0[c]

[a] Decomposition of sample was observed

[b] Data from DSC measurements (Chapter 4.3.5, Table 4.7)

[c] Data from Langmuir-QCM [83]

As one can see from Figure 3.14 the sample condensed on cover has the spectrum significantly different from the spectrum of initial sample. The relative intensities of the IR bands in the region of C-H stretch vibrations are absolutely different in initial sample, residue in the cell after TGA run, and the sample condensed on the cold cover of the device. This change in the intensity is similar to the spectrum changes observed during the 1-butyl-3-methylimidazolium hexafluorophosphate [C$_4$mim][PF$_6$] decomposition in Knudsen experiments [98].

As a matter of fact, the data listed in Table 3.15 for $\Delta_{vap}H^0_{298}$ of the [C$_n$mim][Br] series are actually referred to the complex process of vaporization accompanied with the decomposition of the IL. The measured heat effect $\Delta_{dec}H = \Delta_{vap}H^0$ should be treated as the activation energy of decomposition for [C$_n$mim][Br]. In this case in order to estimate the activation energy of decomposition E_A, the following equation has been applied [98]:

$$\Delta_{dec}H = E_A + 1/2 \cdot RT \qquad (20)$$

According to Eq. (20) activation energies E_A for [C$_n$mim][Br] are: 116.7±1.2 (442.1 K), 115.3±1.8 (432.4 K), 116.3±1.8 (432.4 K) $kJ \cdot mol^{-1}$ for $n=1, 2, 4$, corresponding. For comparison, the [C$_4$mim][PF$_6$] decomposes in a vacuum in the temperature range between 410 to 505 K according to zero-order kinetics with the activation energy $E_A = 68.8\pm2.8$ $kJ \cdot mol^{-1}$ [98]. From this comparison it is apparent that [C$_4$mim][PF$_6$] is less thermally stable than [C$_n$mim][Br]. An important conclusion from studies of the [C$_n$mim][Br] series is that TGA conditions are substantially less mild in comparison to the QCM, where vaporization of ILs is performed in vacuum and, therefore, at lower temperatures. Hence, combination of several methods for study of an IL sample should be a common practice to obtain reliable vaporization enthalpies and to detect possible decomposition processes.

3.4.4 Determination of the Vaporization Enthalpy of [C$_2$mim]$_2$[Co(NCS)$_4$]

Metal-ion-containing ILs are a very interesting subclass of ILs: not only because of the above mentioned properties, but also they may have interesting magnetic or catalytic properties [99, 100]. Investigations on [C$_4$mim][FeCl$_4$] were sensational: it was shown that droplets can be affected by magnets [101, 102, 103]. Therefore, ILs with paramagnetic transition metal cations have been discussed as suitable candidates for magnetic and magnetorheological fluids [104, 105]. In present work we have studied [C$_2$mim]$_2$[Co(NCS)$_4$] (see Fig 3.15) one of the new class of ionic liquids that contain doubly negatively charged tetra isothiocyanatocobaltate (II) anions and that have the appearance of blue ink (see Figure 3.16) [106]. Despite the fact that doubly charged anions are present, surprisingly some of these compounds have glass-transition temperatures that lie far below room temperature and have low viscosity. Furthermore, these ILs are distinguished by useful features, such as good stability towards water and oxygen and also good solubility in many solvents.

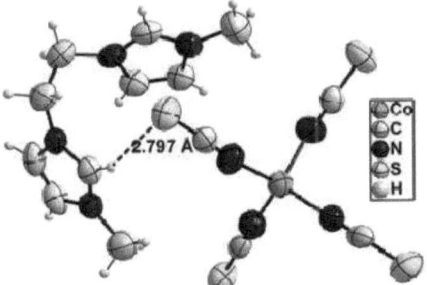

Figure 3.15: Structure of [C$_2$mim]$_2$[Co(NCS)$_4$]

Figure 3.16: Sample of "Blue Ink" – [C$_2$mim]$_2$[Co(NCS)$_4$] with a permanent magnet

Vaporization enthalpy of [C$_2$mim]$_2$[Co(NCS)$_4$] have been measured using the TGA Pyris 6 TGA. Isothermal gravimetric analysis curves were measured in the temperature range 180-240 °C in nitrogen at a flow rate of 140 $mL \cdot min^{-1}$. Mean value of vaporization enthalpy $\Delta_{vap}H_T^0 = 130.9 \pm 0.8$ $kJ \cdot mol^{-1}$ at the average temperature 439.4 K was calculated from three experiments. Extrapolation of the vaporization enthalpy to the reference temperature 298 K has been performed under assumption [72, 73, 74] of the constant difference between heat capacity liquid and gaseous phases of $\Delta_l^g c_p = -105\ J \cdot K^{-1} \cdot mol^{-1}$ and we obtained the value $\Delta_{vap}H_{298}^0 = 150.3 \pm 0.8$ $kJ \cdot mol^{-1}$.

Comparison with a simple half-empirical method of enthalpy of vaporization of [C$_2$mim]$_2$[Co(NCS)$_4$] based on the data on the surface tension (σ) and molar volume (V_m) of the ionic liquids [72]:

$$\Delta_{vap}H_{298} = A \cdot \sigma \cdot V_m^{2/3} \cdot N_A^{1/3} + B \qquad (21)$$

(where N_A is Avogadro's constant, A and B are empirical constants) provides the value of $\Delta_{vap}H_{298}^0 = 287\ kJ \cdot mol^{-1}$. The predicted according equation (21) vaporization enthalpy for the [C$_2$mim]$_2$[Co(NCS)$_4$] is approximately twice larger than those measured in this work $\Delta_{vap}H_{298}^0 = 150.3 \pm 0.8\ kJ \cdot mol^{-1}$. It is apparent that the sizes of the cation and anion in the paramagnetic ionic liquid [C$_2$mim]$_2$[Co(NCS)$_4$] are substantially larger in comparison with common ionic liquid e.g. [C$_2$mim][SCN]. But, surprisingly, the vaporization enthalpy measured in this work for the

[C$_2$mim]$_2$[Co(NCS)$_4$] at 298 K is quite the same as that $\Delta_{vap}H^0_{298}$ = $151\pm2\ kJ \cdot mol^{-1}$ reported [74] for [C2mim][SCN]. Nevertheless, the very low vaporization enthalpy measured in this work is consistent with comparably low viscosity of the [C$_2$mim]$_2$[Co(NCS)$_4$] also measured in this work. Indeed, a number of correlations between vaporization enthalpy of IL with the transport properties: viscosity, self-diffusion coefficients, molar volume have been published recently [107]. For this reason we decided to perform very simple qualitative comparison between vaporization enthalpy of ionic liquids and their viscosity (see Fig 3.17). As can be seen, the low vaporization enthalpy and viscosity of the [C$_2$mim]$_2$[Co(NCS)$_4$] fit very well to the general tendency of the common ILs and this simple comparison provide the necessary mutual validation of the viscosity and vaporization enthalpy studied experimentally in this work.

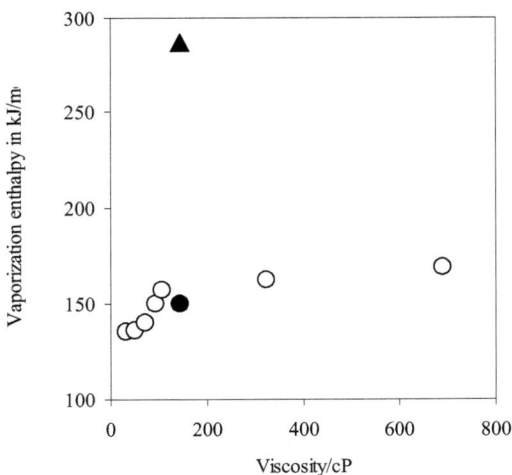

Figure 3.17: Correlation between viscosity of ionic liquids and their vaporization enthalpies. ○ - common ionic liquids [C$_n$mim][NTf$_2$], and [C$_8$mim][BF$_4$], and [C$_8$mim][PF$_6$]; ● - experimental result for vaporization enthalpy of [C$_2$mim]$_2$[Co(NCS)$_4$]; ▲ - vaporization enthalpy of [C$_2$mim]$_2$[Co(NCS)$_4$] calculated according to equation (21)

3.4.5 TGA – outlook and conclusions

Having established the express experimental TGA procedure, an accumulation of vaporization enthalpies of Ils of different structures could be possible. Two main structural ideas could be suggested for the future work. The first one is to fix an anion of Ils under investigation constant and to study an influence of the alkyl chain length of the imidazolium-based [$C_n$mim]-cation. The second idea is to perform systematic studies with the fixed 1-ethyl-3-methyl imidazolium cation ([C_2mim]) and variation of different anions: [Et_2SO_4], [NTf_2], [OTf], [$PF_3(C_2F_5)_3$], [BF_4], [PF_6], [$N(CN)_2$], [$C(CN)_3$], and [SCN]. In addition the following cations could be involved in the study: [C_nPy]; [$C_{1,n}$Pyrr]; [N (C_n)$_4$].

We expect that not in all cases the thermal stability of ILs will allow to obtain the vaporization enthalpy due to possible decomposition. This could be a good indicator to exclude such an IL from studies by other methods, or maybe it will even be a challenging task to study such thermally labile ILs using the Knudsen QCM method, where sensitivity of the sensor and the vacuum conditions could be helpful to avoid decomposition due to low-temperature range for investigation. That is why even simple TGA-screening of ILs vaporization enthalpies with different anions and with different alkyl chains on the cation should add value to the question of feasibility of vaporization studies.

Careful development, attestation and validation of the TGA procedure for determination of vaporization enthalpies of very low volatile compounds have been performed in this work. It is obvious that the TGA results for ILs under study have accuracy of $\pm(3$ to $5)$ $kJ \cdot mol^{-1}$ which can be improved further. Also, further development of this method is indeed, because TGA procedure has three crucial advantages in comparison with the well established methods for vaporization enthalpies measurements: the requirement of relatively small amounts of sample, short actual experimental times, and the simplicity and accessibility of the experimental setup. This methods is highly suitable for the quick screening of the thermal stability range for ionic liquids, combined with the localization of reliable value of vaporization enthalpy, which could be further ascertained by an additional measurements using the traditional techniques such as static and /or Knudsen QCM methods.

4 Introduction to reaction differential scanning calorimetry

Ionic liquids (ILs) exhibit unique physico-chemical properties such as negligibly low vapor pressure at ambient temperatures and thermal stability at elevated temperatures. ILs might be useful as environmentally benign solvents that could replace volatile organic compounds. ILs have also widespread potential for practical applications such as heat transfer and storage medium in solar thermal energy systems as well as for many areas such as fuel cells and rechargeable batteries. By varying the length of the alkyl chains of the cationic core and the type or size of the anion, the solvent properties of ILs can be tailored to meet the requirements of specific applications to create an almost infinite set of "designer solvents". Systematic study of thermodynamic properties within ionic liquids homological series ILs are of large interest, because they are required for the validation and development of the molecular modeling and first-principles methods towards this new class of solvents.

Currently, the most widely used starting material for imidazolium-based ionic liquids are imidazolium halides (mostly bromides and chlorides) that are generally produced in batches as it is typically the case for ionic liquid synthesis. In most cases the IL synthesis reactions are highly exothermic and the kinetics was shown to be fast [108]. The heat management for the reactor design and operation is a crucial point leading to high quality of the resulting product and preventing of thermal runaway. Thus, a systematic accumulation of the data on enthalpies of ILs synthesis reactions is apparently required and differential scanning calorimetry (DSC) is a suitable experimental tool for this purpose.

Differential scanning calorimetry (DSC) is broadly used for studies of heat capacities, enthalpies of phase transition (glass transition, fusion, vaporization) etc. This technique offer important advantages, such as the small quantity of sample necessary for the experiment (a few milligrams), handiness of manipulation, rapidity of performance, versatility and precision of temperature control. DSC is also very useful in the study of physicochemical processes, such as the determination of the kinetics and thermodynamics of polymerization, due to the fact that these type of reactions are exothermic and so they can be easily followed isothermally and dynamically by DSC. This study reports an extention and development of DSC procedure for reliable measurements of enthalpies of ILs synthesis reactions like the following:

$$1 - Me\mathrm{Im}_{liq} + C_nH_{2n+1} - Hal_{liq} \rightarrow [C_nmim][Hal]_{liq} \qquad (22)$$

The thermodynamic properties of imidazolium, pyridinium and pyrrolidinium ILs were derived from the experimental data.

4.1 Experimental section

4.1.1 Materials and chemicals

Samples of 1-methylimidazole (MeIm) and alkyl halides were of commercial origin and they were were distilled under reduced pressure prior to use and stored under dry nitrogen. The degree of purity was determined using a gas chromatograph (GC) equipped with a flame ionization detector. The carrier gas (nitrogen) flow was 7.2 $dm^3 \cdot h^{-1}$. A capillary column HP-5 (stationary phase crosslinked 5% PH ME silicone) was used with a column length of 30 m, an inside diameter of 0.32 mm, and a film thickness of 0.25um. The standard temperature program of the GC was T = 323 K for 180 s followed by heating to T = 523 K with the rate of 10 $K \cdot min^{-1}$. No impurities (greater than 0.05 mass per cent) could be detected in the samples used for the measurements.

4.1.2 Differential Scanning Calorimetry — Measurements of Reaction Enthalpy

The general principle for the heat-flux DSC-measurement of enthalpies of chemical reactions is well documented [109, 110]. In short, in order to obtain heat effect in a dynamic temperature scan of the material the difference between the temperatures of the empty reference pan and sample pan (see Fig. 4.1) is measured and converted to the value of heat flux. The peak corresponding to the heat effect is then integrated over the temperature range of effect. The reaction is considered to be complete when the energy influx or efflux rate levels off to match the baseline. Typical results of a dynamic scan for is shown in Figure 4.2. The total enthalpy of reaction was determined from the surface area of the exothermic peak between the DSC curve and the baseline (see Figure 4.2). The heat released during the IL synthesis reaction is related to the number of mole of alkylhalide available in the starting formulation. The enthalpy of reaction is calculated using the following equation:

$$\Delta_r H_m = \frac{A}{n_r} \frac{1}{1000} \tag{23}$$

where $\Delta_r H_m$ is the enthalpy of reaction, $kJ \cdot mol^{-1}$; A is the peak area, mJ; n_r is the amount of the stoichiometrically deficient reactant, *mmol*. Provided that the careful calibration of the DSC has been done, the integration of the peak area (Fig. 4.2) allows to determine the value of the chemical reaction enthalpy.

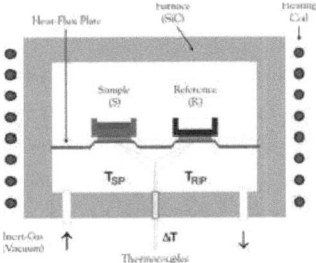

Figure 4.1: Principal scheme of heat-flux DSC apparatus

In the present work, the enthalpy of reaction was measured using the Mettler-Toledo 822 heat flux DSC with a computer-controlled system. Before the measurements on the ILs, two standard materials, indium (99.999% pure) and zinc (99.999% pure), were used to calibrate the temperature and heat flow scales of the DSC.

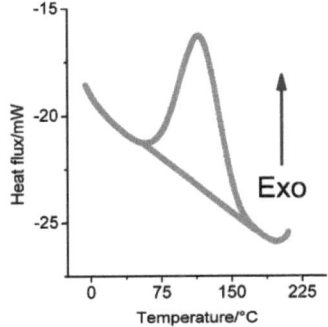

Figure 4.2: A typical peak corresponding to synthesis reaction of an IL

ILs are most commonly synthesized by a quaternization reaction of a substituted amine (or phosphine, N-heterocycles, etc.) with alkyl halides, alkyl sulphates, alkyl-methanesulfonates, etc. As a rule, such N-alkylation reactions are highly exothermic and very fast even at the room temperature. For example, at room temperature the reaction of MeIm with the trifluoromethanesulfonate is extremely fast [111], reaction with the di-ethylsulphate is fast [112], and reaction with butyl chloride is very slow [113]. Two different procedures were applied to prepare the samples for DSC studies, depending on the reaction rate at the room temperature. First one was a common procedure for slow reactions. In the second procedure, some additional precautions were undertaken in order to preclude the reaction begin outside the measuring unit of DSC.

Since the reaction rate at room temperature was considered as negligible, the reactants could be mixed without any precautions. Typically 6-15 mg of starting materials were placed in a standard 40 µL aluminum DSC pan which was weighed with an electronic balance with ±0.000001 g accuracy. Samples masses were reduced to vacuum. Charging of the pan with the starting materials is convenient to perform with 10 µL syringes. The pan was hermetically sealed with a cover to eliminate any evaporation of the reagents. The pan was weighed before and after each experiment to determine any mass loss during the measurements. Experiments with an occasional mass loss were rejected. Measurements were taken at temperatures from 273 K up to 500 K. The scanning rates were usually 10 to 50 $K \cdot min^{-1}$.

Due to the fact that fast reaction release heat already by charging of the pan at the room temperature, the reactants have to be separated inside the pan before the onset of the measurement. A separation membrane or any other mechanical facility is hardly possible to install inside of the tiny (40 µL) aluminum pan. Separation of the reactants in the pan with a wax baffle has heavily aggravated the acquisition of the reaction peak because of an additional fusion enthalpy of the wax. This idea was rejected. Instead, a silicone oil have turned out to be a very practical for reactant separation. Indeed, the silicone oil is chemically inert, thermally stable and as a rule it has negligible miscibility with chemicals common for ionic liquids synthesis. In addition, different types of the commercially available silicone oils have a broad variations of densities. Hence, it is easy to find out a particular silicone oil with the density matching between values of the two starting reactants. Thus, the separation of the reactants in the pan has been adjusted according to the "sandwich" having more dense compound on the bottom of the pan, the silicone oil layer in between, and the second reactant as an upper layer. Heating of the pan inside of DSC leads to a thermal convection and provides an intensive mixing of the reactants. In such a way, a controlled reaction start and heat release have been achieved even for very fast reactions. Preliminary test runs with the pan filled only with the portion of one of reactant together with the silicone oil have confirmed absence of any detectable mixing effects, providing easy acquisition of the reaction peak without taking into account additional heat effects. Thus, the reaction enthalpy ($\Delta_r H_m$) was determined by integration of the experimentally obtained heat flux signal assuming heat effects due to mixing negligible.

For some reactions which slowly start already at room temperature, the reactants could be diluted with the final product to slow down the reaction rate and to avoid an uncontrolled transformation within the stabilization period of the DSC.

4.1.2.1 In solvent or neat?

Commonly, the quaternization reactions are carried out in the batch reactors. Organic solvents (e.g. toluene, ethanol) have often been used for an improved heat removal, increase safety of the process and obtain high quality IL. Many ILs or their intermediates can often be quite viscous. By using a solvent, the product can be kept in a relatively low viscosity liquid solution that can be more easily processed. [114]. However, solvent use leads to increased reaction times due to the dilution and requires quantitative removal in an additional process step.

For our DSC studies use of the solvent has been of crucial importance. In the imidazolium based reactions studied in this work the alkyl-halides were used as alkylating agents. These substances possess very high volatility and the DSC experiments at the neat conditions were hardly possible due to the risk of the pan explosion. Common solvents mentioned above are as rule also volatile. In addition, it has been well established, that e.g. for the synthesis of the ionic liquid 1-butyl-3-methylimidazolium chloride from (MeIm) and 1-chlorobutane already for a MeIm-conversion of 8%, neat synthesis lead to two phases either rich in IL or 1-chlorobutane, whereby MeIm is soluble in both phases. Only an addition of 20 vol.% of ethanol lead to a single-phase synthesis [113]. Such a complex phase behavior during the DSC study could impact the acquisition of the reaction peak. To overcome these two problems, a very low volatile solvent, which is miscible with the reactants and the resulting ionic liquid is required. It has turned out that the most suitable solvent for the imidazolium based ILs studied in this work was another ionic liquid - 1-ethyl-3-methylimidazolium bis-(trifluoromethylsulfono) imide [C_2mim][NTf_2] (99.9%, Iolitec). This solvent was ideal to the decrease the total pressure in the reacting system. It was miscible with the precursors and resulting ILs. In addition, the [C_2mim][NTf_2] as the solvent was very advantageous for a high conversion and selectivity of the MeIm alkylation with alkyl-halides. We tested the influence of amount of [C_2mim][NTf_2] (between 10 wt% and 70 wt%) on the reproducibility of the measured enthalpy of reaction (see Figures 4.3 and 4.4) and selected to keep dilution of the reacting mixture on the optimal level of above 50-70 wt.%.

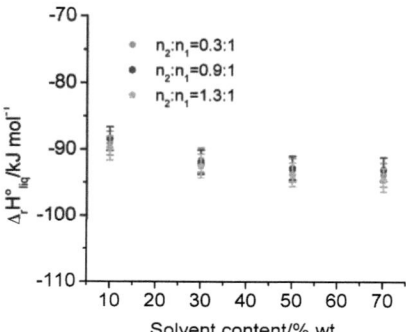

Figure 4.3: Influence of the solvent ($[C_2mim][NTf_2]$) amount on the measured value of enthalpy of formation reaction of $[C_4mim][Br]$

Syntheses of ILs are usually performed in an excess of a highly volatile alkylating reagent in order to reach high degree of conversion and easily purify the prepared IL by gentle heating of an IL in vacuum. However, sometimes reaction was performed at the of MeIm to provide homogeneity of the reaction mixture at the end of the experiment [115]. For our DSC studies of IL synthesis reactions the excess of the volatile component is impractical, because it elevate overload pressure in the pan and the pan usually explodes by heating. In contrast, at the excess of the substituted amine, the pressure inside of the pan remains moderate and it keeps tight during the experiment. From a chemical point of view, the molar reaction enthalpy should be independent on the excess of reactants and it is also apparent from our experimental results given in Figure 4.4.

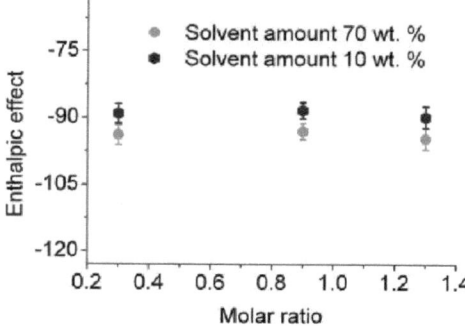

Figure 4.4: The influence of molar ratio on the measured enthalpy of reaction of formation of [C$_4$mim][Br]

4.1.2.3 Temperature Range - Reactions in the Isothermal or in the Scanning Mode?

There are at least four main prerequisites for reliable and reproducible measurements of the reaction enthalpy using DSC:
- high selectivity of chemical reaction;
- completeness of chemical reaction;
- well defined and sharp reaction peak;
- stable baseline with the well defined on-set and end-set of the reaction peak.

Fortunately, the two first requirements - selectivity and high conversion degree, could be generally fulfilled for IL synthesis reactions [115], provided that a reaction is performed at a suitable temperature (or temperature range). Our preliminary experiments (MeIm + BuBr) were carried out under isothermal conditions at temperatures commonly recommended for synthesis of imidazolium based ILs. However, the main problem with isothermal measurements results from the ill-defined behavior of the signal following the initial introduction of the sample. Additionally, in such conditions we obtained very broad reaction peak with an ambiguous conversion of reactants and unclear end-set. Thus, our further experiments were performed under non-isothermal conditions with the dynamic temperature ramp which provided the stable baseline of the DSC-run. We have studied different temperature ramps from of 5 to 50 $K \cdot min^{-1}$ (see Figure 4.5). It has turned out that the most reproducible results could be obtained at scanning rates up 30 $K \cdot min^{-1}$. Thus, in this work the enthalpy measurements were performed at 50 $K \cdot min^{-1}$ heating rate in the temperature range from room temperature up to 530 K.

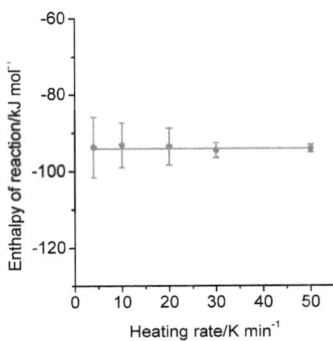

Figure 4.5: The influence of the heating rate on the accuracy of measurement

In general, the selection of the upper temperature of the DSC investigation depends crucially on the thermal stability of ILs under study. For the reactions of imidazolium based halides synthesis (reaction (22) we have deliberately set the end temperature to 523 K in order to obtain the high level of conversion. But it should be mentioned that this end-set temperature was well below the decomposition temperatures (around 560 K) of $[C_n mim][Hal]$ [116, 117]. It is also obvious, that in studies of ILs where decomposition is expected, the upper limit of the DSC study should be adjusted to the lower value which have to be at least of 20-30 K below the decomposition on-set temperature.

In general, the selection of the lowest temperature of the DSC investigation depends on the reaction rate. As a rule, after beginning of the experiment the calorimeter needs certain time to reach the steady state conditions and to stay in this state at least for 2 minutes. It is important that the on-set of the reaction peak is outside of this equilibration period. If the reaction under study is too fast, the lowest temperature of DSC could be selected below the room temperature.

4.1.2.3 Prove of completeness of the IL synthesis reaction

In each experiment we have optimized the temperature range and the scanning rate in order to achieve the maximal possible conversion of the precursor which was in stoichiometric deficiency. In order to check the completeness of the reaction, the aluminum pan with the sample was taken from the DSC and cooled down to the room temperature. After weighing of the pan, its top cover was punctured and the pan was heated at 373-393 K in a special vacuum oven within several hours until the constant mass was reached. In such conditions the IL and the silicone oil could not be evaporated from the pan. Thus the mass loss from the pan was ascribed to the first of the MeIm which was usually taken in excess. Having the record of masses before and after experiment the degree of conversion was calculated according to material balance account. As a rule, the level of conversion in all experiments was between 98% and 99% (with the estimated uncertainty of 0.5%). Small corrections for the incomplete conversion have been taken into account for each reaction under study.

According to results of screening of experimental parameters which has been performed in this work, the following optimal conditions for the DSC studies of the IL synthesis reactions could be recommended:

- ratio of precursors (MeIm):(R-Hal) =1.0:0.3 mol/mol
- content of the solvent – about 70wt. %
- temperature range – 25-250 °C
- heating rate – $50 \ K \cdot min^{-1}$

In these conditions the sharp reaction peak with the well defined on-set, end-set and, baseline could be achieved providing the completeness of chemical reaction.

As a matter of fact, the enthalpy of reaction $\Delta_r H_m$ obtained by integration of the DSC peak is referred to the temperature between the on-set and end-set of the peak. In most cases we observed the symmetrical types of reaction peaks and we could assign the $\Delta_r H_m$ to the temperature of the peak maximum T_{max}. To adjust the enthalpy $\Delta_r H_m$ to standard temperature Kirchgoff's law was used:

$$\Delta_r H_{m,298} = \Delta_r H_{m,T} - \Delta_r c_p \cdot (T - 298.15) \qquad (24)$$

where $\Delta_r c_p$ is change of heat capacity in the system equal to the difference of heat of the product (IL [C$_n$mim][Hal]) and the precursors. The values of heat capacity change for several systems are collected in the Table 4.1.

Table 4.1: Heat capacities of precursors and resulting ILs and heat capacity changes

$c_{p,i}$, $J \cdot K^{-1} \cdot mol^{-1}$			$\Delta_r c_p$, $J \cdot K^{-1} \cdot mol^{-1}$
1-Melm	BuHal	[C$_4$mim][Hal]	
143.3±0.6[121]	BuCl: 158.9±0.6[118]	299.0±7.5[119, 120]	-3.2±8.7
	BuBr: 162.3±0.6[118]	308.7±1.0[122]	3.1±2.2
	BuI: 164.5±4.9[118]	310.0±5.1[123]	2.2±10.6

4.1.3 Computations

Standard *ab initio* molecular orbital calculations were performed using the Gaussian 03 Rev.04 program package [124]. From 20 to 30 initial conformations of ionic liquids were generated which were optimized on B3LYP/6-31+(d,p) level of theory. After that the lowest by energy conformation was calculation on CBS-QB3 level. CBS-QB3 theory uses geometries from B3LYP/6-311G(2d,d,p) calculation, scaled zero-point energies from B3LYP/6-311G(2d,d,p) calculation followed by a series of single-point energy calculations at the MP2/6-311G(3df,2df,2p), MP4(SDQ)/6-31G(d(f),p) and CCSD(T)/6-31Gf levels of theory [125]. Calculated values of the enthalpy and Gibbs energy of ions and ion pairs are based on the electronic energy calculations obtained by the CBS-QB3 and G3MP2 methods using standard procedures of statistical thermodynamics [126].

4.2 Results and discussion

4.2.1 State of the Art

Ionic liquids are often considered as the new class of environmentally-friendly solvents due to their molecularly "tunable" properties and lack of volatility. New applications are being developed at a rapid pace. However, the lack of the kinetic and thermodynamic data of their synthesis with emphasis on reaction engineering and process intensification is acknowledged in the literature [114, 113, 127]. Only a few experimental kinetic and thermochemical studies of the imidazolium halide synthesis reaction (22) were published until December 2010 [114, 113, 127, 128, 129, 111, 130, 131, 121, 122, 123, 132].

Already first kinetic studies of the *[C₄mim][Cl]* synthesis from 1-methylimidazole and 1-chlorobutane in a batch reactor [113, 127, 128]:

$$1 - MeIm\ (liq) + BuCl\ (liq) \rightarrow [C_4mim][Cl]\ (liq) \qquad (25)$$

have shown very complex phase behavior of the reaction participants in the course of reaction [113]. With the MeIm-conversion >8%, neat synthesis lead to two phases either rich in IL or 1-chlorobutane, whereby MeIm is soluble in both phases. Addition of ethanol (>20 vol%) led to a single-phase synthesis and to an increase of the effective rate constant of IL-formation. The reaction was very slow and at 343K about 50 h were needed to reach 90% of MeIm conversion. However, an acceleration of the rate in the neat conditions by increasing the temperature was limited, because of decomposition of MeIm and contamination the IL. Dilution of the reacting system with ethanol as the solvent, ensured a single-phase synthesis and improved the quality of the IL [113].

Schleicher and Scurto [114] performed extended kinetics and solvent effects studies in the synthesis of [C$_6$mim][Br] in 10 different solvents: acetone, acetonitrile, 2-butanone, chlorobenzene, dichloromethane, dimethyl sulfoxide, ethyl formate, ethyl lactate, methanol, and cyclopentanone.

First thermochemical study of the reaction (22) was published by Waterkamp et al. [C4mim][Br] with a reaction enthalpy of -96 $kJ \cdot mol^{-1}$ at 338-358 K and 1-2 bar, but unfortunately, no experimental details available in this work [129]. They also estimated the adiabatic temperature for a run-away reaction for this system and suggested a way to intensify the synthesis of [C4mim][Br] by using a continuously operating microreactor system. Further very promising developments of heat pipe controlled syntheses of ionic liquids in a microchannel reactors were reported by Love et al. [Ill, 130, 131]. The enthalpy of the [C4mim][Br] synthesis reaction:

$$1 - MeIm\,(liq) + BuBr\,(liq) \rightarrow [C_4mim][Br]\,(liq) \quad (26)$$

was determined in a an isoperibol calorimeter $\Delta_r H_{m,298} = -(87.7 \pm 1.6)\ kJ \cdot mol^{-1}$ et al. [129]. However it was mentioned that the enthalpy was determined in a series of five experiments at 334 K. The reaction was performed at the 1.50-1.65 excess of MeIm to provide homogeneity of the reaction mixture at the end of the experiment. The degree of 1-bromobutane conversion at 334 K was only about 95%. In order to obtain enthalpy of reaction according to Eq. 26, the authors [121] have been challenged to take into account:

1. the enthalpy of mixing of 1-bromobutane with MeIm;

2. the enthalpy of dissolution of [C4mim][Br] in IL + MeIm solution;

3. the incomplete conversion of 1-bromobutane;

4. the enthalpy of fusion of [C4mim][Br];

5. adjustment of the reaction enthalpy from 334 K to the reference temperature 298.15 K

The most important steps 1 and 2 were taken into account with a series of calorimetric measurements on model mixtures [121]. The enthalpy of fusion of [C4mim][Br] was obtained from adiabatic calorimetry [122]. Thus, the systematic study [121] of the quaternization reaction according to the Eq 26 seems to provide more reliable value in comparison to ambiguous documented value by Waterkamp et al. [129].

The same working group has recently applied the calorimetric procedures developed in [121] for a determination the enthalpy of the [C$_4$mim][I] synthesis reaction:

$$1 - MeIm\,(liq) + BuI\,(liq) \rightarrow [C_4mim][I]\,(liq) \quad (27)$$

in the isoperibol calorimeter. They obtained $\Delta_r H_{m,298} = -(100.1 \pm 1.2)\ kJ \cdot mol^{-1}$ [123]. The enthalpy of this quaternization reaction was measured in a series of experiments also at 334 K and at the 1.32 to 1.44 excess of MeIm. The measured value was corrected for the side heat effects in the same way is it was described in [121] for the reaction of [C4mim][Br] formation. The standard molar enthalpies of formation for [C4mim][Br] and [C4mim][I] in the crystal and liquid states at T = 298.15 K were determined in [121, 123] from the results for calorimetrically measured $\Delta_r H_{m,298}$.

+

ILs with anions containing chlorine or fluorine (e.g. [PF$_6$]$^-$, [AlCl$_4$]$^-$) are problematic with respect to corrosion and disposal. In homogeneous catalysis the free halogen ions lead to deactivation. Therefore, halogen-free IL synthesis routes were meanwhile developed. An example is the alkylation of methylimidazole with di-ethylsulfate to ethylmethylimidazole di ethylsulfate ([C$_2$mim][EtSO$_4$]), which can be regarded as a bulk IL and the source for various other ILs by subsequent acid-catalyzed trans-esterification with alcohols [132]. Systematic kinetic studies on IL-synthesis according to Eq. (28) were reported in [128, 132]:

$$1 - MeIm\ (liq) + Et_2SO_4\ (liq) \rightarrow [C_4mim][EtSO_4]\ (liq) \tag{28}$$

Meanwhile synthesis of [C4mim][Cl] (see Eq. 25) shows phase separations in solvent-free conditions, the synthesis of [C$_2$mim][EtSO$_4$] occurs in a mono phase [128]. Experimental enthalpy of reaction according to eq (28) $\Delta_r H_{m,298} = -102\ kJ \cdot mol^{-1}$ reported in [128] without any details and this value was used for simulation of heat balance of tubular synthesis reactor. This work was extended and in [127] the kinetic studies on [C$_2$mim][EtSO$_4$] – synthesis were carried out in a continuous tubular reactor. Both reactants were mixed at a low temperature (14°C) to avoid unwanted pre-reactions and the experimental enthalpy of reaction, $\Delta_r H_{m,298} = -102\ kJ \cdot mol^{-1}$ was reported but again without any details [112]. Surprisingly, a significantly higher values of the reaction enthalpy (28) were given in a PhD-Thesis [127] coming from the same working group [128, 112]. Results for $\Delta_r H_{m,298}$ (according to Eq. 28) from three experiments in an adiabatic reaction calorimeter (mixing of 5g MeIm and 5 g Et$_2$SO$_4$) where given as follows : -173 $kJ \cdot mol^{-1}$, -148 $kJ \cdot mol^{-1}$, and -135 $kJ \cdot mol^{-1}$ with the average value $\Delta_r H_{m,298} = -152 \pm 9\ kJ \cdot mol^{-1}$. In addition, an enthalpy of formation of [C$_2$mim][EtSO$_4$] was measured in this work using the static-bomb calorimeter Parr 1341. Only one combustion experiment with 0.5041 g of [C$_2$mim][EtSO$_4$] was performed with the result of molar combustion enthalpy $\Delta_c H_m$

= $-5103\ kJ \cdot mol^{-1}$ and the molar enthalpy of formation $\Delta_f H_m = -621.6\ kJ \cdot mol^{-1}$. However it should be mentioned, that as a rule the combustion experiments with the S-containing compounds are performed with the rotating-bomb combustion calorimeter [133] and not with the static bomb as reported in [127].

Renken et al. [108] has also reported kinetic studies and the use of a microstrue-tured reactor system for the production of [C$_2$mim][EtSO$_4$] by a solvent-free alkylation reaction. The reaction kinetics was studied by DSC at temperatures 293-443 K. Experiments were carried out under isothermal as well as non-isothermal conditions by imposing a well-defined temperature ramp to the samples. They observed that reaction starts already at room temperature. Therefore, the reactants were diluted with the final product to slow down the rate, thus reducing the error due to uncontrolled transformation within the stabilization period of the DSC. The reaction enthalpy has been determined from calorimetric experiments with reactants diluted with the final product. A dilution of a factor 10 was found to be necessary to get reproducible results of $\Delta_r H_m = -190 \pm 2\ kJ \cdot mol^{-1}$. This result is in line with published values [113] but in the disagreement with results from adiabatic calorimetry [127].

Minnich et al. [134] combined reaction calorimetry with ATR-IR spectroscopic monitoring of kinetics and thermodynamics of [C$_2$mim][EtSO$_4$] synthesis reaction. Semi-batch solvent-free experiments were performed in a reaction calorimeter (Mettler Toledo) under an inert atmosphere (argon) at 308 K. One reactant was charged to the reactor, while the second was thermally pre-equilibrated and injected via a thermostated dosing unit with slight overpressure of argon. Integration of the discrete heat flow pulses yielded the reaction enthalpies per injection. Values were correlated with the degree of conversion that has been determined gravimetrically. Surprisingly, $\Delta_r H_m$ was not independent from conversion. The absolute values of $\Delta_r H_m$ showed a systematic trend with conversion including an increase to nearly the 1.5-fold of the starting exothermicity and a decrease at conversions higher than 0.9. For the synthesis where MeIm was charged and di-ethylsulfate was added, the range of the measured reaction enthalpies was between -120 and -145 $kJ \cdot mol^{-1}$. For the synthesis where di-ethylsulfate was charged and MeIm was added, the range of the measured reaction enthalpies was generally lower - between -85 and -140 $kJ \cdot mol^{-1}$. The averaged molar reaction enthalpy of the N-alkylation of 1-methylimidazole with di-ethylsulfate was found to be between -123 and -130 $kJ \cdot mol^{-1}$. This range of $\Delta_r H_m$-values is apparently closer to the result from adiabatic calorimetry [127].

To sum up: available in the literature calorimetric results for reaction enthalpies of the IL synthesis reactions from precursors discussed in chapters 4.2.1.1 and 4.2.1.2 are in disarray.

Complex phase behavior during the synthesis, large enthalpic effects and diverse reaction rates have hardly aggravated the reliability of the measured values. Disagreement among the data reported for the same reaction in literature sources sometimes exceed 50 $kJ \cdot mol^{-1}$ Development of an accurate and reproducible calorimetric procedure is highly desired.

4.2.3 *Molar Enthalpies of Formation of ILs from DSC-Measurements*

Differential scanning calorimetry (in comparison to the adiabatic reaction calorimetry) is a simple and reliable experimental technique which allows examining thermal properties of different substances, among them are thermal effects of different phase transitions. Obviously the idea to apply DSC for the investigation of thermochemistry of ILs appears to be very attractive. The systematic study of parameters crucial for the reliable DSC-experiment described in chapter *2* have demonstrated that DSC could be successfully used to determine reaction enthalpies of the synthetic step of IL from precursors. Certainly, these $\Delta_r H_m$ – values are practically important for the heat management, assessment of thermal runaway and quality of product. However, at the same time the $\Delta_r H_m$ results are also of theoretical value to derive molar enthalpies of formation of ILs $\Delta_f H_m$ from DSC-Measurements. As it will be shown below, the enthalpies of formation of ionic liquids in combination with the high-level *ab initio* calculations open a new way to obtain enthalpies of vaporization of ILs [73, 135, 136].

Traditionally, the combustion calorimetry is used as the well established method to provide enthalpies of formation [73]. However, this method is sophisticated and needs extended experiences to obtain reliable results. Also this method has several shortages and restrictions. The most crucial prerequisite for the combustion calorimetry is the quality and amount of sample required for the study. In order to complete the experiment successfully, about 10-20 g of highly pure (99.9%) sample are required. Such amount of ILs is very expensive and also methods for purity attestations of the sample are still not sufficiently developed. Another issue is that in order to get an acceptable accuracy of the results (with an uncertainty of 1-3 $kJ \cdot mol^{-1}$ in $\Delta_r H_m$) the combustion energy have to be measured with an error of 0.01-0.02%. Such result is already achieved for C, H, O, and N containing compounds using the static bomb calorimeter [73, 137]. For the ILs additionally containing Cl, Br, I, and S atoms, special procedure with the rotating bomb calorimeter is required. This procedure has now succeeded for the S-containing ILs [138] and further design for the Cl, Br, and I containing ILs is in progress in our laboratory. In this context we put our efforts for the development of the new DSC based thermochemical option leading to the values of the enthalpies of formation and further, indirectly, to the vaporization enthalpies.

As a representative example, we consider once more the reaction according to Eq. 22 to show the way how the enthalpy of reaction of synthesis of [C4mim][Cl] is used to obtain its enthalpy of formation. According to the Hess law the enthalpy of reaction is calculated as follows:

$$\Delta_r H^0_{liq} = \Delta_f H^0_{liq}([C_4 mim][Cl]) - \Delta_f H^0_{liq}(MeIm) - \Delta_f H^0_{liq}(BuCl) \quad (29)$$

where $\Delta_r H^0_{liq} = -(78.7 \pm 3.0)\ kJ \cdot mol^{-1}$ was measured in this work using DSC in the liquid phase; $\Delta_f H^0_{liq}(MeIm) = (70.1 \pm 1.0)\ kJ \cdot mol^{-1}$ [139]; $\Delta_f H^0_{liq}(BuCl) = -(188.1 \pm 1.2)\ kJ \cdot mol^{-1}$ [140].

From Eq. (29) the enthalpy of formation of the IL can be calculated:

$$\Delta_f H^0_{liq}([C_4 mim][Cl]) = -78.7 + 70.1 - 188.2 = -196.7 \pm 5.2\ kJ \cdot mol^{-1}$$

In the same way the enthalpies of formation of other ILs [C$_n$mim][Hal] were estimated (see Table 4.5). These values can be used (with help of results from *ab initio* calculations) to obtain the enthalpies of vaporization of these ILs indirectly (see Chapter *3.3.1*).

Table 4.3: Experimental results for measurements of reaction enthalpies $\Delta_r H^0_q$ using DSC.

System (1) + (2)	Molar ratio n_2 / n_1	Solvent content, % wt.	$\Delta_r H^0_{liq}$, kJ·mol^{-1}	Peak temperature, °C
Imidazolium				
MeIm + BuCl	0.30	75	-78.7±3.0	230±8
MeIm + PenCl	0.25	75	-76.4±3.0	225±6
MeIm + HexCl	0.19	75	-77.4±2.8	220±10
MeIm + HepCl	0.17	80	-75.8±3.4	220±5
MeIm + OctCl	0.15	80	-78.0±3.7	220±5
MeIm + BuBr	0.40	70	-89.0±1.5	175±5
MeIm + PenBr	0.44	70	-94.6±2.6	160±5
MeIm + HexBr	0.32	70	-89.4±3.1	160±2
MeIm + HepBr	0.36	75	-87.5±2.8	160±2
MeIm + OctBr	0.30	75	-89.2±3.2	185±5
MeIm + BuI	0.67	0	-100.4±0.9	135±2
MeIm + Et$_2$SO$_4$	1.38	85	-151.3±1.5	140±5
Pyridinium				
Pyridine + BuBr	0.43	85	-87.3±1.4	166±1
Pyridine + PenBr	0.30	80	-87.1±2.0	165±2
Pyridine + HexBr	0.28	85	-87.5±2.0	166±5
Pyridine + HepBr	0.50	90	-87.4±2.4	166±3
Pyridine + OctBr	0.38	90	-87.3±2.2	165±3
Pyrrolidinium				
N-MePyrr+BuBr	0.43	95	-79.2±1.9	117±3
N-MePyrr+PenBr	0.43	95	-79.1±2.0	120±4

N-MePyrr+HexBr	0.53	95	-81.5±2.0	118±2
N-MePyrr+HepBr	1.10	90	-77.7±2.6	117±3
N-MePyrr+ OctBr	1.51	95	-76.9±2.4	118±4

Table 4.5: Enthalpies of formation of ILs $\Delta_f H^0_{q\,IL}$ derived from reaction enthalpies

System (1)+(2)	$\Delta_f H^0_{liq,2}$, kJ·mol^{-1}	$\Delta_r H^0_m$, kJ·mol^{-1}	$\Delta_f H^0_{liq,IL}$, kJ·mol^{-1}
Methylimidazolium, $\Delta_f H^0_{liq,1}$ = 70.1 ± 1.0 kJ·mol^{-1} [139]			
MeIm + BuCl	-188.2±1.2 [140]	-78.7 ± 3.0	-196.8±5.2
MeIm + PenCl	-213.4±1.3 [140]	-76.4 ± 3.0	-219.7±5.3
MeIm + HexCl	-239.4±1.5 ^{0}	-77.4 ± 2.8	-216.7±5.3
MeIm + Hep Cl	-264.8±1.6 ^{0}	-75.8 ± 3.4	-270.5±6.0
MeIm + OctCl	-291.3±1.3 [140]	-78.0 ± 3.7	-299.2±6.6
MeIm + BuBr	-144.2±1.3 [141]	-89.0 ± 1.5	-163.1±3.8
MeIm + PenBr	-170.4±1.5 [141]	-94.6 ± 2.6	-194.9±5.1
MeIm + HexBr	-194.5±1.6 [141]	-89.4 ± 3.1	-213.8±5.7
MeIm + HepBr	-218.6±1.6 [141]	-87.5 ± 2.8	-236.0±5.4
MeIm + OctBr	-215.2±2.3 [141]	-89.2 ± 3.2	-264.3±6.3
MeIm + Et$_2$SO$_4$	-812.9±1.0 [142]	-151.3 ± 1.5	-894.1±3.5
Pyridine, $\Delta_f H^0_{liq,1}$ = 100.0 ± 0.5 kJ·mol^{-1} [143]			
Pyridine + BuBr	-144.2±1.3	-87.3 ± 1.4	-131.5±3.2
Pyridine+PenBr	-170.4±1.5	-87.1 ± 2.0	-157.5±4.0
Pyridine+HexBr	-194.5±1.6	-87.5 ± 2.0	-182.0±4.1
Pyridine+HepBr	-218.6±1.6	-87.4 ± 2.4	-206.0±4.5
Pyridine+OctBr	-2 15.2±2.3	-87.3 ± 2.2	-232.5±5.0

4.2.4 Indirect Determination of Molar Enthalpies of Vaporization of ILs

Enthalpies of vaporization should be intensively measured, as these data are of a paramount importance for the proper calibration and validation of new force fields used in computational simulations based on MD technique, as well as for anchoring parameters for p-V-T equations of state for neat ILs and mixtures. Since the first quantitative measurements of vapor pressure and vaporization enthalpies of the [C$_n$mim][NTf$_2$] published in 2006 [72, 17] only twelve new papers dealing with these thermodynamic properties have appeared in the literature [85, 76, 71, 88, 72, 73, 74, 135, 144, 145, 146]. Up to date, only 27 values of the vaporization enthalpies of ILs have been measured using different techniques. Most of ILs studied are the imidazolium based [C$_n$mim]+ with the anions: [NTf$_2$], [BF$_4$], [EtSO$_4$], [PF$_6$], [N(CN)$_2$], [SCN], [FeCl$_4$], and [CF$_3$SO$_3$]. In addition, there are data of two [C$_n$Py]$^+$ with the anions: [NTf$_2$], [BF$_4$], and also for [C$_{n,n}$Pyrr]$^+$ with the anions: [PF$_3$(C$_2$F$_5$)$_3$] and [N(CN)$_2$]. About 2/3 of all available data were measured using the temperature programmed desorption (TPD) combined with mass spectrometry (MS) [76, 74, 145].

It should be mentioned, that this technique is quite new and has never been applied before for quantitative measurements of vaporization enthalpies. Vaporization of IL from the open surface in a hot zone occurs in this method. There are still some difficulties in the MS-signals interpretation [144] and there are some uncertainties of data treatment according to the Langmuir open surface vaporization procedure. In spite of these complications, the results from TPD MS method are in very good agreement with the results from well established Knudsen method [72] and transpiration method [135]. A similar idea as TPD MS have been exploited recently: instead of the MS, the conjunction with single photon ionization by a tunable vacuum ultraviolet synchrotron source have been used [146]. Apparently, due to very sophisticated procedure of the data treatment, the uncertainties in vaporization enthalpy associated with this method were of the level of ±12 kJ · mo/$^{-1}$. Nevertheless, the available data on vaporization enthalpies of ILs are rather a minefield than a consistent collection of data. For example, the data for the most frequently studied homologues series of [C$_n$mim][NTf2] is presented on the Fig. 3.9. The disagreement among results from different techniques is typically on the level of 5-7 kJ · mo/$^{-1}$, but for the data from the direct drop-calorimetric study [88] disagreement reaches even 3040 kJ · mo/$^{-1}$. This fact shows clearly the necessity of a more conclusive investigation of the vaporization enthalpies of ILs using reliable experimental techniques. At the same time, any indirect option for obtaining enthalpies of vaporization from empirical, half-empirical or theoretical methods is of a great value to provide consistency of the data.

4.2.5 Vaporization Enthalpy from Enthalpies of Formation $\Delta_f H_{liq}^0 (IL)$

One of the first indirect options to determine enthalpy of vaporization is the combination of combustion calorimetry with the high-level quantum chemical calculations. It has turned out to be a valuable thermochemical option to derive vaporization enthalpies [73, 135, 136]. In the recent time we have developed a new alternative procedure to obtain vaporization enthalpies. This procedure is based on the experimental results from combustion calorimetry and the first-principles calculations according to the general equation:

$$\Delta_f H_g^0 = \Delta_f H_{liq}^0 + \Delta_{vap} H \tag{30}$$

where $\Delta_f H_{liq}^0 (IL)$ - is measured using combustion calorimetry; $\Delta_f H_g^0 (IL)$ - *is* calculated by first-principles methods (e.g. G3MP2 method).

Enthalpy of vaporization of the IL can be evaluated from eq. (30):

$$\Delta_{vap}H = \Delta_f H_g^0 - \Delta_f H_{liq}^0 \tag{31}$$

A comprehensive studies of the thermochemical properties of [C$_n$mim][N(CN)2], [C$_n$mim][NO$_3$], and [C$_{1;4}$Pyrr][N(CN)$_2$] have been performed successfully for validation of this procedure [73, 135, 136, 137, 138, 144, 147].

In this work we have extended this procedure for combination of the high-level quantum chemical calculations with the results from DSC to derive vaporization enthalpies. This procedure is based on the experimental on $\Delta_f H_{liq}^0 (IL)$ from DSC and the first-principles calculations according to the general equation. As an example - for the ionic liquid [C$_4$mim][Cl] using the following data: $\Delta_f H_{liq}^0$ = —196.7±3.4 kJ· mo/$^{-1}$ - is derived from DSC (see chapter 3.3.2); $\Delta_f H_g^0$ = —43.5 kJ · mo/$^{-1}$ - is calculated by first-principles method (CBS-QB3 method); the enthalpy of vaporization of the IL could be estimated from eq. (31):

$$\Delta_{vap}H = -43.5 - (-196.7) = 153.2 \pm 3.4 \; kJ \cdot mol^{-1} \tag{32}$$

This indirect value is in a good agreement with the result $\Delta_{vap}H$ = 153.5 ± 2.3 obtained from TGA (see Table 3.15). Values of $\Delta_f H_{liq}^0$ of ILs under study (see Table 4.5) together with results from CBS-Q3B calculations have been used to obtain the enthalpies of vaporization for these ILs indirectly (see Table 4.7).

4.2.6 Vaporization Enthalpy from DSC Enthalpies of Reaction, $\Delta_r H_m^0$

The indirect option to determine enthalpy of vaporization using Eq. (31) suggested in 3.3.4 has one handicap: it is based on the value of the gaseous enthalpy of formation $\Delta_f H_g^0$ which is evaluated using *ab initio* calculations. There are some ambiguities in respect to procedures of converting results of first-principles calculations (total energies E$_0$ at T = 0 K and enthalpies H$_{298}$ at T = 298.15 K) into the desired value of $\Delta_f H_g^0$. It is well established that in standard Gaussian-n theories, theoretical standard enthalpies of formation $\Delta_f H_g^0$ are calculated through atomization reactions, isodesmic and bond separation reactions [148]. However, it has turned out [148, 149] that gaseous enthalpies of formation calculated from the standard atomization procedure deviate very often systematically from the available experimental results. The deviation could be positive or negative, depending on the homologous series under study [149]. It is also often discussed, that the enthalpies of formation obtained from the first-principle calculations are very sensitive for the choice of the bond separation or isodesmic reactions used for this purpose [148]. The possible reasons are usually explained for different quality of the experimental data, involved in estimation, as well as a disbalance of electronic energies calculated for the reaction participants. That is why we are somewhat reticent to use of atomization reactions, isodesmic and bond separation reactions to obtain $\Delta_f H_g^0$ of ionic liquids. At the same time the logic of the DSC experiment lend itself an elegant solution of the aforementioned problem. Indeed, using DSC we have measured the experimental reaction enthalpy in the liquid phase e.g. for [C$_4$mim][Cl] according to Eq. (25).

$$\Delta_r H_{exp,liq}^0 = \Delta_f H_{liq}^0 ([C_4mim][Cl]) - \Delta_f H_{liq}^0 (MeIm) - \Delta_f H_{liq}^0 (BuCl) = -78.7 \pm 3.0 \ kJ \cdot mol^{-1}$$

The same reaction enthalpy, but in the gaseous phase could be obtained from the first-principles calculation according to Eq. (33):

$$MeIm(g) + 1 - BuCl(g) \rightarrow [C_4mim][Cl](g) \qquad (33)$$

We have calculated using the CBS-Q3B the enthalpies H$_{298}$ at T = 298.15 K for all reaction participant of Eq. (33). According to the Hess Law the enthalpy of this reaction is calculated as follows:

$$\Delta_r H_g^0 = H_{298,g} ([C_4mim][Cl]) - H_{298,g} (MeIm) - H_{298,g} (BuCl) = -14.9 \ kJ \cdot mol^{-1} \quad (34)$$

It is important to emphasize that the $\Delta_r H_g^0$ is obtained using first principles directly from the calculated H$_{298}$ by-passing the conventional calculation of the enthalpies of formation with help of atomization or isodesmic procedures. As a consequence of this trick, the resulting value of $\Delta_r H_g^0$ is not affected at all by choice and quality of auxiliary quantities commonly required for atomization or bond separation reactions.

Having established the values of $\Delta_r H_{exp,liq}^0$ and $\Delta_r H_g^0$ we have derived the vaporization enthalpy (e.g for [C$_4$mim][Cl]) provided that the following balance holds:

$$\Delta_f H_{liq}^0 = \Delta_f H_g^0 - \Delta_{vap} H \qquad (35)$$

The enthalpy of reaction was calculated using the Hess law:

$$\Delta_r H_{exp,liq}^0 = [\Delta_f H_g^0 (IL) - \Delta_{vap} H(IL)] - [\Delta_f H_g^0 (MeIm) - \Delta_{vap} H(MeIm)] - [\Delta_f H_g^0 (BuCl) - \Delta_{vap} H(BuCl)] \qquad (36)$$

The enthalpy of vaporization of the IL has been obtained from the following equation:

$$\Delta_{vap}H(IL) = -\Delta_r H^0_{exp,liq} + \Delta_r H^0_g + \Delta_{vap}H(MeIm) + \Delta_{vap}H(BuCl) \quad (37)$$

With the following values of vaporization enthalpies for reactants:

$\Delta_{vap}H(MeIm) = 55.1 \pm 0.2\ kJ \cdot mol^{-1}$ [139] and $\Delta_{vap}H(BuCl) = 33.5 \pm 0.1\ kJ \cdot mol^{-1}$ [150] the Eq. 37 gives:

$\Delta_{vap}([C_4mim][Cl]) = -(-78.7) + (-14.9) + 55.1 + 33.5 = 152.4 \pm 3.3\ kJ \cdot mol^{-1}$

So, using the value $\Delta_r H^0_g$ calculated with CBS-Q3B in the gas phase and enthalpies of vaporization of 1-Methyl-Imidazole and alkyl halides available from the literature [139, 150], and the enthalpy of reaction $\Delta_r H^0_{exp,liq}$ measured using DSC in the liquid phase, the enthalpies of vaporization of the ILs under study have been derived (see Table 4.7). For comparison, the value $\Delta_{vap}H([C_4mim][Br]) = 156.8 \pm 2.4\ kJ \cdot mol^{-1}$ has been determined from DSC (Table 4.7) being in a very good agreement with the result from the Quartz-Crystal-Microbalance method [83]: $\Delta_{vap}H([C_4mim][Br]) = 157.9 \pm 3.0\ kJ \cdot mol^{-1}$. This test result demonstrates in a convincing way the application of the described DSC method to "indirectly" obtain vaporization enthalpies of ILs.

Comparison of the vaporization enthalpies derived using two procedures: from Eq. (31) and from Eq. (37) is given in Table 4.7. Both methods provide the values in a close agreement, however we should recommend the Eq. (37) because the $\Delta_r H^0_g$ for this procedure have been calculated directly from the H_{298} avoiding the atomization procedure.

Table 4.7: Comparison of vaporization enthalpies derived using Eq. (31) and Eq. (37) (in $kJ \cdot mol^{-1}$).

IL	$\Delta_f H^0_{liq,exp}$	$\Delta_f H^0_{g,CBS}$	$\Delta_{vap}H_{298}$ From Eq. (31)	$\Delta_{vap}H_{298}$ From Eq. (37)
1	2	3	4	5
Chlorides				
[C$_1$mim][Cl]	-101.3	31.8	133.1	136.3
[C$_2$mim][Cl]	-138.7	0.4	139.1	140.8
[C$_3$mim][Cl]	-165.5	-22.9	142.6	144.7
[C$_4$mim][Cl]	-196.7	-43.5	153.2	152.4
[C$_5$mim][Cl]	-219.5	-63.9	155.6	154.5
[C$_6$mim][Cl]	-246.0	-84.1	161.9	160.0
[C$_7$mim][Cl]	-270.0	-105.4	164.6	161.8
[C$_8$mim][Cl]	-299.2	-126.9	172.3	167.2
Bromides				
[C$_1$mim][Br]	-68.3	68.0	136.3	141.4

[C$_2$mim][Br]	-107.0	33.6	140.6	145.9
[C$_3$mim][Br]	-141.7	10.6	152.3	149.5
[C$_4$mim][Br]	-167.2	-10.5	156.8	156.8
[C$_5$mim][Br]	-194.7	-32.5	162.2	160.5
[C$_6$mim][Br]	-213.5	-51.5	162.0	161.2
[C$_7$mim][Br]	-235.8	-71.5	164.3	164.1
[C$_8$mim][Br]	-264.2	-91.6	172.6	170.8

4.3 Structure — property relationships for thermodynamic properties of ionic liquids. Correlation with the number of C-atoms in the cation alkyl chain

The physical and chemical properties of a compound are a function of its molecular structure. Structure-property relationships are easy to follow for homologous series of organic compounds. Thermodynamic properties (enthalpies of formation, enthalpies of vaporization, etc.) have been often correlated with the chain length, retention indices, topological indices, molecular descriptors [151]. A relationship, once quantified, can be used to estimate the properties of other molecules simply from their structures and without the need for experimental determination or synthesis. This has resulted in the development of quantitative structure-property relationships (QSPR) [152] or group-additivity procedures [153, 154] as an important tool in chemical, biological, and environmental research. Group-additivity (GA) methods are well recognized approaches to prove consistency of the experimental data on thermochemical properties of molecular compounds [153, 154]. One of the best indicators of possible experimental errors is a large deviation between experimental values and those calculated by GA - especially if other, closely related compounds show no such discrepancy. The most successful GA method for estimating of thermodynamic properties was suggested by Benson [153, 154]. This method serves as a valuable tool for many scientists and engineers whose work involves thermodynamic characterization of elementary and overall reaction processes.

The correlations derived from experimental datasets can be used to predict the physical properties of related molecules directly from their modeled structures [155]. When a structure-property relationship is found, it may also provide insight into which aspect of the molecular structure influences the property. Such insight can facilitate a systematic approach to the design of new molecules with more desirable properties.

Unfortunately, for ionic liquids the available thermodynamic data are still scarce and the possibilities for prediction these properties are very restricted yet [147, 137]. In this context the structure - property relationships for thermodynamic results derived in this work could serve rather for testing of internal consistency of the data as well as for validation of the DSC procedure developed in this work. There at least two obvious ways to test the data for ILs for internal consistency.

A first one is a ***correlation of a thermodynamic property with the number of C-atoms in the cation alkyl chain***. For example for ILs of the general formula [C$_n$mim][X], the thermodynamic property (vaporization enthalpy, formation enthalpy etc.) is expected to increase monotonically in accordance with the increase in the CH$_2$

The second way is only a modification of the first way and it has been referred to study of ***application of group-additivity rules for ILs***. Are GA methods able to predict thermochemical properties of ILs properly? Getting a definitive answer to this question is difficult. The available information is too limited for a quantitative parameterization of any groups specific for ILs. A further question arises – is it generally possible to transfer group contributions established for common organic compounds to ionic liquids? Due to scarcity of the available data, only simple structural patterns could be considered before the development of any kind of GA method. It has turned out that the imidazolium based ionic liquids of general formula [C$_n$mim][Anion] with n= 2 and 4 have been studied in the literature most frequently. We collected the thermochemical data on [C$_n$mim][Anion] available from the literature and from this study in Table 3.9.

In order to get insight into possible group additivity rules for ionic liquids we have compared differences in enthalpies (vaporization or formation) of structurally parent molecular liquids and ionic liquids.

$$CH_3\text{-}[CH_2]_2\text{-}CH_3 \text{ and } CH_3\text{-}[CH_2]_4\text{-}CH_3$$
$$CH_3\text{-}[CH_2]_2\text{-}OH \text{ and } CH_3\text{-}[CH_2]_4\text{-}OH$$
$$[C_2\text{mim}][X] \text{ and } [C_4\text{mim}][X]$$

These differences could serve as simple indicators of whether or not molecules obey additivity rules. We have used both ways for testing of internal consistency of the thermodynamic data derived using the DSC procedure developed in this work.

4.3.1 Enthalpy of IL synthesis reaction, $\Delta_r H_m^0$

The correlation of enthalpies of IL synthesis reactions, $\Delta_r H_m^0$ (see Table 4.3) with the number of C-atoms in the series of homologues with a cations of different alkyl chain length is a valuable test to check the internal consistency of the experimental results. Do we expect the linear dependence $\Delta_r H_m^0$ on the chain length? Before we discuss this question regarding the ionic liquids, let us consider the general tendencies in the molecular compounds - alkyl halides, which are reaction participants. On the one hand, it is well established, that vaporization enthalpies appear to be a linear function of the number of carbon atoms of the alkyl chain in aliphatic esters [156] and aliphatic nitriles [157]. The plot of $\Delta_{vap} H_{298}$ against the number of C-atoms in the linear aliphatic halides $CH_3-(CH_2)_n-Hal$ with n = (2 to 15) is presented in Figure 4.6. On the other hand, as can be seen in Figure 4.6, the first two representatives - methyl and ethyl halides are slightly out of the linear correlation. Such an anomaly has been also observed for aliphatic esters [156] and aliphatic nitriles [157] and this fact might be caused by the high dipole moment of these species The dependence of vaporization enthalpy on the total number of C-atoms for $N_c > 2$ is expressed by the following equations:

$$\Delta_{vap}H_{298}(AlkCl), kJ \cdot mol^{-1} = 14.3 + 4.8 \cdot N_c \quad (r^2 = 0.9995) \quad (38)$$

$$\Delta_{vap}H_{298}(AlkBr), kJ \cdot mol^{-1} = 22.5 + 4.8 \cdot N_c \quad (r^2 = 0.9996) \quad (39)$$

from which enthalpies of vaporization of other representatives of these two series with $N_c > 2$ can be calculated

It is also well known that formation enthalpies $\Delta_f H^0_{g,298}$ follow the linear function of the number of carbon atoms of the alkyl chain in different homologous series of the organic compounds [158]. The plot of $\Delta_f H^0_{g,298}$ against the total number of C-atoms in the linear aliphatic halides $CH_3-(CH_2)_n-Hal$ with n = (2 to 15) is presented in Figure 4.7. However, for this property the first two representatives - methyl and ethyl halides are again slightly out of the linear correlation (see Figure 4.7). The dependence of enthalpy of formation on the total number of C-atoms for $N_c > 2$ is expressed by the following equations:

$$\Delta_f H^0_{m,g,298}(AlkCl), kJ \cdot mol^{-1} = -85.0 - 25.7 \cdot N_c \quad (r^2 = 0.9995) \quad (40)$$

$$\Delta_f H^0_{m,g,298}(AlkBr), kJ \cdot mol^{-1} = -43.5 - 25.1 \cdot N_c \quad (r^2 = 0.9996) \quad (41)$$

from which enthalpies of formation $\Delta_f H^0_{m,g,298}$ of other representatives of these two $N_c > 2$.

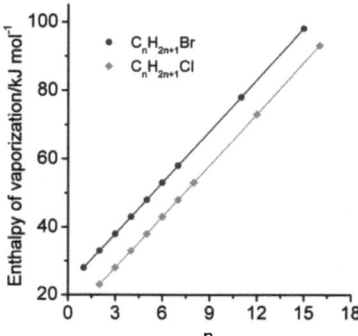

Figure 4.6: Enthalpies of vaporization $\Delta_{vap}H_{298}$ of alkylhalides of various alkyl lengths [158].

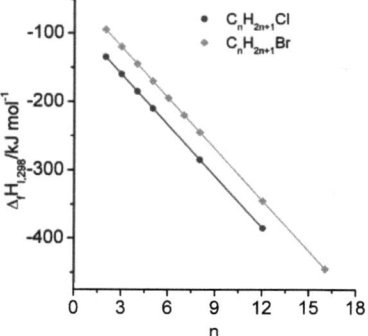

Figure 4.7: Enthalpies of formation $\Delta_f H^0_{g,298}$ of alkylhalides of various alkyl lengths [158].

Coming back to the correlation of enthalpies of IL synthesis reactions $\Delta_r H^0_m$ (see Table 4.5) with the N_c – number in the cation alkyl chain, it is apparent (see Figure 4.8), that enthalpies calculated (using the CBS-Q3B) for reactions of formation of [C$_n$mim][Hal] according to Eq. (33) show the similar behavior as discussed above. Indeed, the reactions enthalpies with the methyl and ethyl halides are noticeably less negative. However, beginning from N$_c$ > 4, the calculated $\Delta_r H^0_{m,g}$ are practically independent on the chain length. The same trend could be observed for $\Delta_r H^0_{m,liq}$ measured in this work (see Table 4.3) using the DSC. Unfortunately, methyl, ethyl, and propyl halides are too volatile two perform the DSC studies properly. That is why we studied in this work only reactions (see Figure 4.8) with CH$_3$ — (CH$_2$)$_n$ — Ha/ with n = (3 to 7). However, as it can be seen from Figure 4.8, the measured $\Delta_r H^0_{m,liq}$ values for imidazolium, pyridinium and pyrrolodinium based ILs are most likely independent on the chain length (larger than butyl) within the boundaries of the experimental uncertainties. Since, the experimental and calculated $\Delta_r H^0_{m,g}$ and $\Delta_r H^0_{m,liq}$ have shown the similar tendencies, we have assumed the same trend in $\Delta_r H^0_{m,liq}$ for [C$_1$mim][Hal], [C$_2$mim][Hal], and [C$_3$mim][Hal], where the experiments were not feasible. This assumption has allowed to estimate enthalpies of vaporization of for [[C$_1$mim][Hal], [C$_2$mim][Hal], and [C$_3$mim][Hal] in the same way as it was done in Chapter 4.2.5.

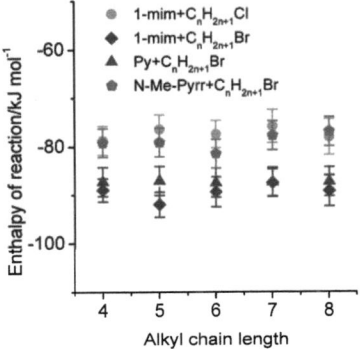

Figure 4.8: Enthalpies of reactions in liquid phase of 1-methylimidazole, pyridine, and N-methyl-pyrrolidine with alkylhalides of various alkyl lengths (Data from Table 4.3).

4.3.2 *Enthalpies of vaporization of IL, $\Delta_{vap}H_{298}$*

It is well established, that enthalpies of vaporization in most homologous series C$_n$H$_{2n+1}$ — R (alkanes [87], 1-substitued-alkanes [87], alcohols [159], esters [156], aldehydes [160] etc.) linearly correlate with the chain length N_c. Does enthalpy of vaporization of ILs correlate with the number of C-atoms in the cation alkyl chain linearly as it has been shown for vaporization enthalpies of alkyl halides in Chapter 3.4.1? The answer on to this question is still evasive. For example, for ILs of the general formula [$C_n$mim][NTf$_2$] with n= 2, 4, 6, and 8, the vaporization enthalpy is expected to increase monotonically from [C_2mim][NTf$_2$] to [C_8mim][NTf$_2$] in accordance with the with the increase in the number of CH$_2$ groups (the contribution to vaporization enthalpy of 5.0 $kJ·mol^{-1}$ per each CH$_2$ group is typical for the n-alkane family [161]). However, this is in disagreement with experimental observations available from the literature as shown in Table 3.7. Indeed, the enthalpies of vaporization of the ethyl and the butyl derivatives should be different by about 10 $kJ · mol^{-1}$, but they are hardly distinguishable within the boundaries of their experimental uncertainties. The enthalpy of vaporization of the hexyl derivate is expected to be again by 10 $kJ · mol^{-1}$ larger than those of the previous [C_4mim][NTf$_2$], however the experimental value is only slightly larger (see Table 3.7). Only between [C_6mim][NTf$_2$] and [C_8mim][NTf$_2$] the assumed contribution of 5.0 $kJ · mol^{-1}$ each CH$_2$ group comes true first. The reason for such an anomaly (or non-additivity) in the behavior of ILs could be due to interplay of the van der Waals and Coulombic interactions. The increase of the van der Waals interactions with the chain length (as observed for pure n-alkanes), assumed for the ILs, is virtually impacted with the by electrostatic forces and the energy required for the transfer of the contact ionic pair to the gaseous state [15].

Using the vaporization enthalpies of [$C_n$mim][Hal] estimated using the DSC (Table 4.7) we are able to add the value to the discussion of linearity of these ILs. The he effect of n-alkyl chain length on the vaporization enthalpies was probed using the data from Table 4.7.

As can be seen on the Figure 4.7, the vaporization enthalpy is increasing rather monotonically from C_1 to C_8 for both homological series under study: chlorides and bromides. Some spread of the values around the linear tendency is quite acceptable taking into account the experimental uncertainties of the $\Delta_r H^0_{m,liq}$ being typically on the level of (± 2-3 $kJ · mol^{-1}$). Thus, the linear correlation for vaporization enthalpies of ILs on the cation alkyl chain length has been verified in this work using the data indirectly derived from DSC measurements.

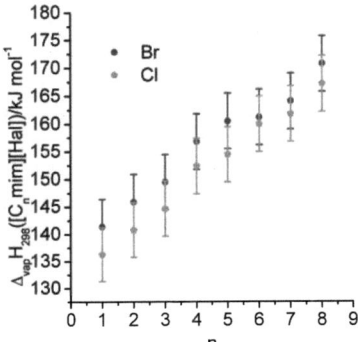

Figure 4.9: Enthalpies of vaporization of [C$_n$mim][Hal] and alkylhalides of various alkyl lengths (Data from Table 4.7).

In the literature there are only about 30 available data points of vaporization enthalpies [17, 74, 76, 85, 145, 146] In order to compare differences in enthalpies vaporization of structurally parent molecules:

$$CH_3-[CH_2]_2-CH_3 \text{ and } CH_3-[CH_2]_4-CH_3$$
$$CH_3-[CH_2]_2-OH \text{ and } CH_3-[CH_2]_4-OH$$
$$[C_2mim][X] \text{ and } [C_4mim][X]$$

Table 4.9: Compilation of the enthalpies of vaporization $\Delta_{vap}H_{298}$, enthalpies of formation in the liquid $\Delta_f H^0_{m,liq,298}$, and in the gaseous phase $\Delta_f H^0_{m,g,298}$ some alkanes, alcohols, and imidazolium based ILs (in $kJ \cdot mol^{-1}$).

Compound	$\Delta_{vap}H_{298}$	$\Delta_f H^0_{m,liq,298}$	$\Delta_f H^0_{m,g,298}$
1	2	3	4
CH$_3$—[CH$_2$]$_2$—CH$_3$[87]	22.4±0.1	-146.6±0.7	-125.6±0.7
CH$_3$—[CH$_2$]$_4$—CH$_3$[a][87]	31.7±0.1	-198.7±0.8	-167.1±0.8
Δ[a]	**9.3**	**-52.1**	**-41.5**
CH$_3$—[CH$_2$]$_2$—OH[159]	52.3±0.1	-302.6±0.5	-255.1±0.5
CH$_3$—[CH2]$_4$—OH[159]	61.7±0.1	-351.6±0.4	-294.7±0.5
Δ[a]	**9.4**	**-49.0**	**-39.6**
[C$_2$mim][Cl][b]	140.8	-138.7	2.1[c]
[C$_4$mim][Cl][b]	152.4	-196.7	-44.3[c]

Δ^a	11.6	-58.0	-46.4
$[C_2mim][Br]^b$	145.9	-107.0	38.9^c
$[C_4mim][Br]^b$	156.8	-167.2	-10.4^c
Δ^a	10.9	-60.2	-49.3
$[C_2mim][NTf_2]^d$	123.1±1.0	—	—
$[C_4mim][NTf_2]^d$	129.7±1.0	—	—
Δ^a	6.6		
$[C_2mim][N(CN)_2]$	157.0±2.1	235.3±3.1	392.3±3.7
$[C_4mim][N(CN)_2]$ [73]	157.2±1.1	195.0±2.7	352.2±2.9
Δ^a	0.2	-40.0	-40.1
$[C_2mim][NO_3]$ [71]	163.7±5.3	-216.9±2.0	-53.2±4.9
$[C_4mim][NO_3]$ [71]	162.1±5.7	-261.4±2.9	-99.0±4.9
Δ^a	-1.3	-44.5	-45.8

^{a} - Difference between respective values for C_4 and C_2 derivatives. ^{b} - Values from the Table 4.7. ^{c} - Sum of the columns 2 and 3. ^{d} - Data from QCM method (Table 3.13)

The experimental data on vaporization enthalpies of two n-alkanes and n-alcohols are presented in column 2 of Table 4.9. We discuss these differences to understand whether or not ILs obey additivity rules. It is evident from this column that the differences between $CH_3 — [CH_2]_4 — X$ and $CH_3 — [CH_2]_2 — X$ *are* identical (9.3 $kJ \cdot mol^{-1}$) within the boundaries of the experimental uncertainties in alkanes and alcohols. We should expect a similar tendency for ILs. Indeed, the corresponding differences between $[C_4mim][X]$ and $[C_2mim][X]$ for halides and $[NTf_2]$ (Table 4.9, column 2) are close to 9.3 $kJ \cdot mol^{-1}$ (within the boundaries of the experimental uncertainties) and the additivity for this property seems to exist for IL and the vaporization enthalpy. However, the data for $[C_nmim][X]$ with $X = [N(CN)_2]$ and $[NO_3]$ do not fit this conclusion, e.g differences between $[C_4mim][X]$ and $[C_2mim][X]$ for both anions (Table 4.9, column 2) are close to zero. One of the possible reasons could be outlying of the first representatives of the homological series with the $[Anion] = [N(CN)_2]$ or $[NO_3]$ as for alkyl halides. We can conclude that may be the group additivity rules for enthalpies of vaporization of some series of $[C_nmim][X]$ ionic liquids could be valid for the ILs with alkyl chain length from $N_c > 2$.

4.3.3   Enthalpies of formation of ILs in the liquid state, $\Delta_f H^0_{m,liq}$

Enthalpies of formation in most homologous series $C_nH_{2n+1} — X$ of organic compounds usually linearly correlate with the chain length $N_c>2$ [158]

The plot of $\Delta_f H^0_{m,liq,298}$ against the number of C-atoms in series $[C_nmim][Hal]$ from Table 4.5 is presented in Figure 4.10. As can be seen in Figure 4.10, the first representative $[C_1mim][Hal]$ is slightly out of linear correlation. Such an anomaly has been also observed for enthalpies of formation of alkyl halides above. The dependence of enthalpy of formation on the total number of C-atoms for $N_c > 2$ is expressed by the following equations:

$$\Delta_f H^0_{m,liq,298}([C_n mim][Cl]), kJ\cdot mol^{-1} = -87.3 - 26.4\cdot N_c \quad (r^2 = 0.999) \quad (42)$$

$$\Delta_f H^0_{m,liq,298}([C_n mim][Br]), kJ\cdot mol^{-1} = -63.0 - 25.2\cdot N_c \quad (r^2 = 0.994) \quad (43)$$

from which enthalpies of formation $\Delta_f H^0_{m,liq,298}$ of other representatives of these $Nc > 2$

Experimental data on enthalpies of formation $\Delta_f H^0_{m,liq,298}$ of two n-alkanes and n-alcohols are presented in column 3 of Table 4.9. The difference between CH$_3$ — [CH$_2$]$_4$ — X and CH$_3$ — [CH$_2$]$_2$ — X is approximately -50 kJ· mol^{-1} for both homologous series. The corresponding difference for the [C$_2$mim][X] and [C$_4$mim][X] (Table 4.9, column 3) is between -(40÷60) kJ· mol^{-1} which is acceptable (within the boundaries of experimental uncertainties) agreement with $2\times(CH_2) = 2\times(-25.5) = -51.0$ kJ · mol^{-1} expected [154] from the molecular compounds. However, due to the relatively high combined experimental uncertainties of about ±(3÷4) kJ · mol^{-1} more experimental information is still required to establish additivity rules for $\Delta_f H^0_{m,liq,298}$ of ILs.

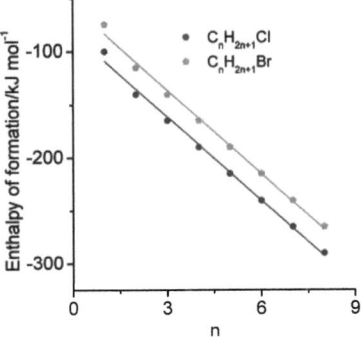

Figure 4.10: Enthalpies of formation $\Delta_f H^0_{m,liq,298}$ of [C$_n$mim][Hal] of various alkyl lengths (see Table 4.5).

4.3.4 Enthalpies of formation of ILs in the gaseous state, $\Delta_f H^0_{m,g,298}$

For enthalpies of formation of ILs in the gaseous phase $\Delta_f H^0_{m,g,298}$ the similar correlations with the chain length N_c as for the $\Delta_f H^0_{m,liq,298}$ have been observed. The plot of $\Delta_f H^0_{m,g,298}$ against the number of C-atoms in series [C$_n$mim][Hal] from Table 4.7 is presented in Fig. 4.11.

Again, the first representative – $[C_1mim][Hal]$ is slightly out of the linear correlation. The dependence of enthalpy of formation on the total number of C-atoms for $N_c > 2$ is expressed by the following equations:

$$\Delta_f H^0_{m,g,298}([C_n mim][Cl]), kJ \cdot mol^{-1} = 41.2 - 21.0 \cdot N_c \quad (r^2 = 0.9997) \quad (44)$$

$$\Delta_f H^0_{m,g,298}([C_n mim][Br]), kJ \cdot mol^{-1} = 73.2 - 20.7 \cdot N_c \quad (r^2 = 0.9992) \quad (45)$$

from which enthalpies of formation $\Delta_f H^0_{m,g,298}$ of other representatives of these two $N_c > 2$.

Experimental data on enthalpies of formation $\Delta_f H^0_{m,g,298}$ of two n-alkanes and n-alcohols are presented in column 4 of Table 4.9. The differences between $CH_3 — [CH_2]_4 — X$ and $CH_3 — [CH_2]_2 — X$ as well as for $[C_4mim][X]$ and $[C_2mim][X]$ (Table 4.9, column 4) seem to be remarkably consistent. Analysis of differences between $[C_4mim][X]$ and $[C_2mim][X]$ (see column 4, Table 4.9) reveals that the contribution from two CH_2 groups in ionic liquids seems to be somewhat more negative (an average of about -45 $kJ \cdot mol^{-1}$) in comparison to the value $2 \times (CH2) = 2 \times (—20.9) = —41.8\ kJ \cdot mol^{-1}$ which is expected for normal molecular compounds. However, this difference for ionic liquids could be considered as acceptable, taking into account the uncertainties assigned for the CBS-Q3B method of ± 5.0 $kJ \cdot mol^{-1}$. We conclude that group additivity rules for the gaseous enthalpies of formation of the series of $[C_n mim][X]$ ionic liquids are valid for the ILs with both short and long alkyl chain lengths on the cation. However, more experimental information is required for detecting similarities or disparities of additivity parameters of molecular and ionic compounds.

The accumulation of data on vaporization enthalpies and enthalpies of formation of ionic liquids is essential for development of another kinds of correlations between structures of ionic liquids and their thermodynamic properties.

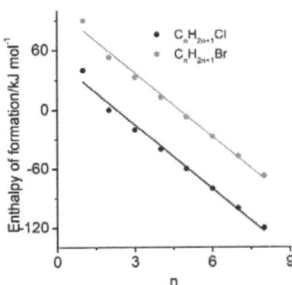

Figure 4.11: Enthalpies of formation $\Delta_f H^0_{m,g,298}$ of [C$_n$mim][Hal] of various alkyl lengths (see Table 4.7).

4.4 Calculation of the Gas Phase Enthalpies of Reaction and Formation. Quantum Chemical Calculations

Ab initio calculations of ILs under study were performed using the Gaussian 03 Rev.04 program package [124]. Initially the conformational analysis of the cations of ionic liquids was carried out. The butyl derivatives of ionic liquids containing C_4 top were used as models. Rotational conformers of 1-n-butyl-3-methylimidazolium, N – butylpyridinium and N – methyl – N - butylpyrrolidinium cation were studied at the level RHF/3-21G* at 0 K. Molecular structures and relative energies of all conformations for the cations formed by rotation of alkyl groups around N-C and C-C bonds of the butyl group (CCCN and CCNC dihedrals) by 360° have been studied with 10° step starting from the co-planar conformation (Fig. 40 for 1-n-butyl-3-methylimidazolium). The conformers with the lowest energies were chosen for further calculations (Fig. 41 and Table 18). For C_2 — C_3, C_5 — C_8 cations the model conformers of C_4 were increased or decreased for appropriate carbon atoms. The optimization of the obtained structures was followed by the analysis of the Mulliken charges distributions. The halogen anion was placed near the atoms with the highest positive charge. In this way from 10 to 20 conformers of ionic liquids were generated, which were optimized on B3LYP/6-31+(d,p) level of theory.

After this conformers generation the lowest by energy conformer was calculated on CBS-QB3 level. CBS-QB3 theory uses geometries from B3LYP/6-311G(2d,d,p) calculation, scaled zero-point energies from B3LYP/6-311G(2d,d,p) calculation followed by a series of single-point energy calculations at the MP2/6-311G(3df,2df,2p), MP4(SDQ)/6-31G(d(f),p) and CCSD(T)/6-31Gf levels of theory [125]. Calculated values of the enthalpy and Gibbs energy of ions and ion pairs are based on the electronic energy calculations obtained by the CBS-QB3 and G3MP2 methods using standard procedures of statistical thermodynamics [126].

4.4.1 Enthalpy of formation $\Delta_f H^0_{m,g,298}$ - improvement of the atomization procedure

We have calculated using the CBS-QB3 a total energies E_0 at T = 0 K and enthalpies H_{298} at T = 298 K. In the Gaussian-n theories, theoretical standard enthalpies of formation $\Delta_f H^0_{m,g,298}$ are calculated through the atomization reactions or bond separation reactions [162]. Raghavachari et al. [163] have proposed to use a set of isodesmic reactions, the "bond separation reactions" to derive theoretical enthalpies of formation. Isodesmic reactions conserve the number of types of bonds and should thus be an improvement of simple atomization reactions. Further enhancement in the calculation of enthalpies of formation should be provided by homodesmic reactions, which, in addition to the types of bonds, also conserve the hybridization of the atoms in the bond. Unfortunately, for ionic liquids the choice of species with the reliable thermodynamic data involved in the isodesmic or bond separation is very restricted. In this work we have applied the atomization procedure (AT), e.g for [C$_2$mim][Cl]:

$$[C_2 mim][Cl] \rightarrow 6 \cdot C + 11 \cdot H + 2 \cdot N + Cl \qquad (46)$$

In contrast to the bond separation and isodesmic reactions, the atomization reaction suggests the constant quality of the data on the right side (the constituted atoms with the well established thermodynamic values) the same for all compounds under study [148].

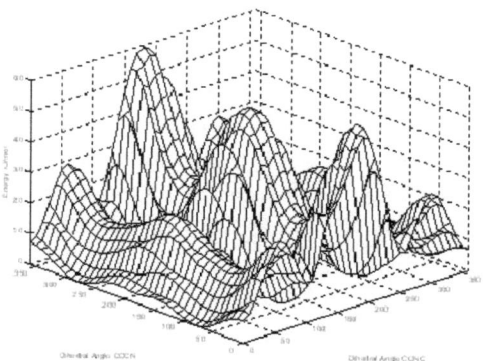

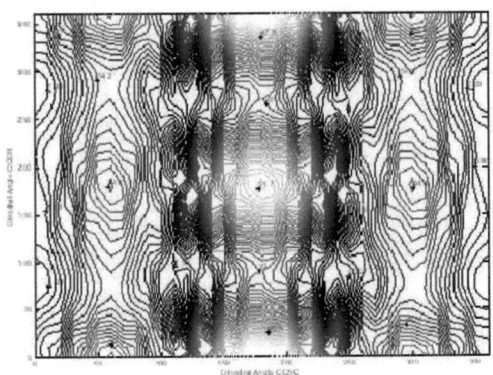

Figure 4.12: Potential surface of internal rotation butyl group in n-butyl-3-methylimidazolium cation.

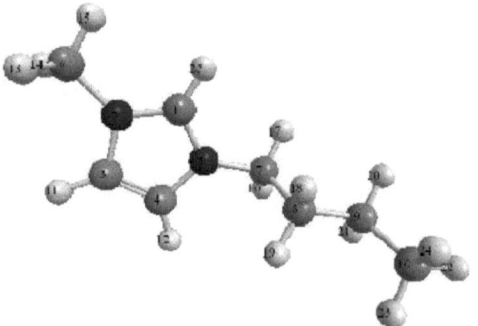

Figure 4.13: The conformers with the lowest energies of n-butyl-3-methylimidazolium cation.

Table 4.11: Geometrical parameters of the most stable conformers of 1-n-butyl-3-methylimidazolium cation, B3LYP/6-31G(d)

Parameter	Value	Parameter	Value
C(1)-N(2)	1.339	H(23)-C(10)-H(24)	107.905
C(1)-N(5)	1.338	N(5)-C(1)-N(2)-C(3)	0.129
C(1)-H(25)	1.080	N(5)-C(1)-N(2)-C(6)	-179.610
N(2)-C(3)	1.383	H(25)-C(1)-N(2)-C(3)	179.821

N(2)-C(6)	1.470	H(25)-C(1)-N(2)-C(6)	0.082
C(3)-C(4)	1.364	N(2)-C(1)-N(5)-C(4)	-0.175
C(3)-H(11)	1.079	N(2)-C(1)-N(5)-C(7)	-178.271
C(4)-N(5)	1.383	H(25)-C(1)-N(5)-C(4)	-179.867
C(4)-H(12)	1.079	H(25)-C(1)-N(5)-C(7)	2.037
N(5)-C(7)	1.483	C(1)-N(2)-C(3)-C(4)	-0.032
C(6)-H(13)	1.092	C(1)-N(2)-C(3)-H(11)	-179.671
C(6)-H(14)	1.092	C(6)-N(2)-C(3)-C(4)	179.707
C(6)-H(15)	1.090	C(6)-N(2)-C(3)-H(11)	0.068
C(7)-C(8)	1.532	C(1)-N(2)-C(6)-H(13)	-120.009
C(7)-H(16)	1.094	C(1)-N(2)-C(6)-H(14)	119.330
C(7)-H(17)	1.093	C(1)-N(2)-C(6)-H(15)	-0.347
C(8)-C(9)	1.537	C(3)-N(2)-C(6)-H(13)	60.296
C(8)-H(18)	1.098	C(3)-N(2)-C(6)-H(14)	-60.364

However, before application of the AT procedure for estimation of enthalpies of formation $\Delta_f H^0_{m,g,298}$ of ILs it is reasonable to test validity of the AT method to reproduce the reliable experimental data properly. In Table 4.13 and 4.15 we collected two sets of reliable enthalpies of formation for chlorine- and bromine-containing compounds. Results of calculations using the CBS-QB3 method are given in the column 4, Table 4.13 and 4.15. It is apparent that enthalpies of formation calculated from the atomization procedure are *systematically* less than those from the experimental data for all chlorine- and bromine-compounds under the study. Notwithstanding, as it has been mentioned above, the enthalpies of formation from the atomization procedure deviate systematically from the experimental values. It has turned out that a remarkable linear correlations could be found between experimental enthalpies of formation and those calculated by atomization enthalpies of formation:

$$\Delta_f H^{0,exp}_{m,g,298}(AlkCl), kJ \cdot mol^{-1} = 3.5 + 0.998 \cdot \Delta_f H^{0,AT}_{m,g,298}(AlkCl) \quad (r^2 = 0.9996) \quad (47)$$

$$\Delta_f H^{0,exp}_{m,g,298}(AlkBr), kJ \cdot mol^{-1} = 28.2 + 1.002 \cdot \Delta_f H^{0,AT}_{m,g,298}(AlkBr) \quad (r^2 = 0.9981) \quad (48)$$

Using this correlation, the "corrected" enthalpies of formation of the compounds under study have been calculated (see Tables 4.13-4.15) and these values are now in good agreement with the experimental results. Having established this "modified atomization procedure" we could use the eq. (47) and (48) for estimation enthalpies of the Cl- and Br-containing compounds from the CBS-QB3 energies without creating any "suitable" bond separation or isodesmic reactions. Thus, the CBS-QB3 method combined with the modified atomization procedure have been applied in this work for reliable calculations of $\Delta_f H^0_{m,g,298}$ of the ILs (see Tables 4.17-4.19).

Table 4.13: Results of Calculation of the Enthalpy of formation Chlorine-containing Compounds in the Gaseous Phase at 298.15 K, CBS-QB3, $kJ \cdot mol^{-1}$

Compound	$\Delta_f H_{m,g,298}^{0,exp}$	$\Delta_f H_{m,g,298}^{0,calc}$	$\Delta = \Delta_f H_{m,g,298}^{0,exp} - \Delta_f H_{m,g,298}^{0,calc}$
Chloromethane	-81.9±0.5	-87.4	5.5
1-Chloroetane	-112.1±1.1	-116.1	4.0
1-Chloropropane	-131.9±1.0	-136.2	4.3
2-Chloropropane	-144.9±1.0	-150.5	5.6
1-Chlorobutane	-154.4±1.3	-156.0	1.6
Chloroethylene	37.2 : 1.2	18.9	18.3*
2-Chloro-propen-1	-21.0±9.4	-22.9	1.9
Chlorocyclohexane	-163.7±3.6	-167.1	3.4
Chlorobenzene	52.0±1.3	48.2	3.8
di-Chloromethane	-95.4±1.1	-105.5	10.1*
tri-Chloromethane	-102.7±1.2	-120.6	17.9*
tetra-Chloromethane	-95.7±1.0	-125.4	29.7*

Table 4.15: Results of Calculation of the Enthalpy of formation Bromine-containing Compounds in the Gaseous Phase at 298.15 K, CBS-QB3, $kJ \cdot mol^{-1}$

Compound	$\Delta_f H_{m,g,298}^{0,exp}$	$\Delta_f H_{m,g,298}^{0,calc}$	$\Delta = \Delta_f H_{m,g,298}^{0,exp} - \Delta_f H_{m,g,298}^{0,calc}$
2-Bromopropane	-98.5	-132.9	24.7
Bromomethane	-37.5	-68.2	30.7
Hydrogen bromide	-36.4	-61.7	25.3
Bromoethane	-63.6	-94.1	30.5
Bromocyan	186.2	160.9	25.3
Bromoethene	79.2	51.3	27.9
Br_2	30.9	-8.3	39.2
Bromoisopropane	-99.4	-123.1	23.7
Bromopropane	-87.0	-113.9	26.9
Bromopropene	45.2	11.4	33.9
Bromobenzene	105.4	82.0	23.4
Bromobutane	-107.1	-134.0	26.9
Bromopentane	-129.0	-160.4	31.4
2-Bromoisobutane	-132.4	-156.9	24.5
Bromoethene	79.2	51.3	27.9

Table 4.17: Results of Calculation of the Enthalpy of formation Chlorine-containing ILs in the Gaseous Phase at 298.15 K, CBS-QB3, $kJ \cdot mol^{-1}$

IL	G3MP2	CBS-QB3	
		$\Delta_f H_{m,g,298}^{0,AT}$	$\Delta_f H_{m,g,298}^{0,AT}$ corr
[C_1mim][Cl]	44.1	28.3	31.8
[C_2mim][Cl]	10.2	-3.2	0.4
[C_3mim][Cl]	-15.1	-26.5	-22.9
[C_4mim][Cl]	-37.5	-47.2	-43.5
[C_5mim][Cl]	-59.4	-67.6	-63.9
[C_6mim][Cl]	-81.0	-87.8	-84.1
[C_7mim][Cl]		-109.2	-105.4
[C_8mim][Cl]		-130.7	-126.9
[C_1Py][Cl]	92.7	77.5	80.9

IL			
[C₂Py][Cl]	54.5	52.6	56.0
[CsPy][Cl]	27.3	27.0	30.5
[C₄Py][Cl]	4.4	5.9	9.4
[C₅Py][Cl]	-17.7	-15.1	-11.5
[C₆Py][Cl]	-39.4	-35.4	-31.8
[C₇Py][Cl]		-55.4	-51.7
[C₈Py][Cl]		-75.4	-71.7
[C₁,₁Pyrr][Cl]	—	-108.5	-104.7
[C₁,₂Pyrr][Cl]	-128.3	-134.4	-130.6
[C₁,₃Pyrr][Cl]	—	-156.6	-152.7
[C₁,₄Pyrr][Cl]	-173.6	-177.2	-173.3
[C₁,₅Pyrr][Cl]	-195.4	-197.6	-193.6
[C₁,₆Pyrr][Cl]	-216.9	-217.8	-213.8
[C₁,₇Pyrr][Cl]		-237.9	-233.8
[C₁,₈Pyrr][Cl]		-257.9	-253.8

Table 4.19: Results of Calculation of the Enthalpy of formation Bromine-containing ILs in the Gaseous Phase at 298.15 K, CBS-QB3, $kJ \cdot mol^{-1}$

IL	CBS-QB3	
	$\Delta_f H^{0,AT}_{m,g,298}$	$\Delta_f H^{0,AT}_{m,g,298}$ corr
[C₁mim][Br]	39.7	68.0
[C₂mim][Br]	5.4	33.6
[C₃mim][Br]	-17.6	10.6
[C₄mim][Br]	-38.6	-10.5
[C₅mim][Br]	-60.6	-32.5
[C₆mim][Br]	-79.6	-51.5
[C₇mim][Br]	-99.6	-71.5
[C₈mim][Br]	-119.6	-91.6
[C₁Py][Br]	84.9	113.3
[C₂Py][Br]	56.6	84.9
[C₃Py][Br]	39.4	67.7
[C₄Py][Br]	13.0	41.2
[C₅Py][Br]	-1.9	26.3
[C₆Py][Br]	-23.5	4.7
[C₇Py][Br]	-42.9	-14.8
[C₈Py][Br]	-62.3	-34.2
[C₁,₁Pyrr][Br]	-98.3	-70.2
[C₁,₂Pyrr][Br]	-126.3	-98.3
[C₁,₃Pyrr][Br]	-146.2	-118.2
[C₁,₄Pyrr][Br]	-166.8	-138.9
[C₁,₅Pyrr][Br]	-188.7	-160.8
[C₁,₆Pyrr][Br]	-208.3	-180.4
[C₁,₇Pyrr][Br]	-228.1	-200.2
[C₁,₈Pyrr][Br]	-247.8	-220.0

4.4.2 Quantum chemical calculations for degree of dissociation of the ion pair

In recent papers [73, 135, 149] we have shown, that the aprotic ILs such as [C$_4$mim][N(CN)$_2$], [Pyrr$_{1,4}$][N(CN)$_2$], and [C$_n$mim][NO$_3$] exist in the gaseous phase as contact ion pairs and not as separated ions. In this work calculations for the free cation [C$_2$mim]$^+$, [C$_2$Py]$^+$, [C$_{1,2}$Pyrr]$^+$, the free anions [Cl]$^-$, [Br]$^-$, and the ion pairs [C$_2$mim]$^+$[Cl]$^-$, [C$_2$mim]$^+$[Br]$^-$, [C$_2$Py]$^+$[Cl]$^-$, [C$_2$Py]$^+$[Br]$^-$, [C$_{1,2}$Pyrr]$^+$[Cl]$^-$, [C$_{1,2}$Pyrr]$^+$[Br]$^-$ have been performed. Results of the electronic energy at 0 K, the molecular structures in the lowest energetic state and all vibrational frequencies of each species have been obtained. On the basis of these quantum mechanical results, the molar Gibbs energy, the molar enthalpy, and the molar entropy at 298.15 K have been calculated using standard procedures of statistical thermodynamics. The purpose of this procedure was to obtain values of the molar standard Gibbs energy of reaction $\Delta_r G°$, the standard molar entropy of reaction $\Delta_r S°$, and the molar reaction enthalpy $\Delta_r H°$ for the process of dissociation of the ion pair, e.g [C$_2$mim]$^+$[Cl]$^-$ into ions in the gaseous phase according to the following equation:

$$[C_2 mim]^+[Cl]^- \rightarrow [C_2 mim]^+ + [Cl]^- \tag{49}$$

The chemical equilibrium constant K_p in the ideal gaseous state has been calculated at 298.15 K using the CBS-QB3 method according to the equation:

$$K_p = \exp\left(-\frac{\Delta G}{RT}\right) \tag{50}$$

from which the degree of dissociation of the ion pair a defined as:

$$\alpha = \frac{p^+}{P_{IP} + p^+} = \frac{p^-}{P_{IP} + p^-} \tag{51}$$

can be calculated according to:

$$K_p = \frac{\alpha^2}{1-\alpha^2} p_{sat} \tag{52}$$

$$\alpha = \sqrt{\frac{K_p}{K_p + p_{sat}}} \tag{53}$$

where $p^+ = p^-$ is the partial pressure of the cation and anion respectively; P_{IP} is the partial pressure of the ion pair; p_{sat} - saturated pressure of the IL ($p_{sat} = P_{IP} + p^+ + p^-$).

We have obtained $Kp = 1.4 \cdot 10^{-65}$ for [C$_2$mim]$^+$[Cl]$^-$ and the equilibrium constant for dissociation of [C$_2$mim]$^+$[Br]$^-$, [C$_2$py]$^+$[Cl]$^-$, [C$_2$py]$^+$[Br]$^-$, [C$_{1,2}$pyrr]$^+$[Cl]$^-$, [C$_{1,2}$pyrr]$^+$[Br]$^-$, have been on the same level. These values are in agreement with

$Kp = 1.4 \cdot 10^{-54}$ for [C$_4$mim][N(CN)$_2$] derived in our previous work [73]. Such extremely low values allow us to conclude with high reliability that the degree of dissociation of the ion pair is zero for the ionic liquids under study, i.e. these ionic ILs exist exclusively as ion pairs in the gaseous phase.

4.4.3 Quantum chemical calculations for degree of conversion of the IL synthesis reaction

One of the most important prerequisites for the successful studies of the IL synthesis reactions by DSC is the complete conversion of the starting material into the IL. Using the first principles calculation it is possible to assess the possible yield of ILs. For example, for the typical reaction according to Eq. (54):

$$C_nH_{2n+1}Hal + 1-MeIm \rightarrow [C_nmim][Hal] \qquad (54)$$

the equilibrium constants for such chemical reactions in the gaseous phase K_P, could be calculated using any suitable method. Using the value of K_P, it is easy to estimate the equilibrium constant in the liquid phase K_A which is required to assess the yield:

$$K_A = K_P \cdot \frac{p^0_{R-Hal} \cdot p^0_{MeIm}}{p^0_{IL}} \qquad (55)$$

where p^0_i are the saturated vapor pressures of the pure reactants. Saturated vapor pressures for alkyl halides [164] and Me-Imidazole [82] are available in literature. For the IL we have deliberately but arbitrary applied an unrealistically high value of p^0_{IL} = 0.001 Pa at 298.15 K in order to embrace the general trends. Using values of K_P for the gaseous IL synthesis reactions calculated with help of CBS-QB3, G3MP2 or DFT methods, the true thermodynamic constants K_A according to Eq. (55) in the liquid phase were calculated. It has turned out that the calculated constants K_A for the model reactions were in the range from 30 to $2.9 \cdot 10^{13}$. Following the chemical equilibrium of the reactions under study are totally shifted to the right side (independent of the method applied for estimations) and for these reactions the very high yields of ILs are expected in agreement with the experimental results from the DSC measurements. Thus, the useful procedure have been developed in this work which could be practically applied for assessment of yields of IL synthesis reactions.

4.4.4 Enthalpy of the IL synthesis reaction. Composite Methods

In this work we have applied two composite methods: CBS-QB3 and G3MP2 to estimate $\Delta_r H^0_{g,IL}$. G3MP2 method is less time consuming in comparison to CBS-QB3, but G3MP2 is also less precise. For this reason all $\Delta_r H^0_{g,IL}$ have been calculated using the CBS-QB3 method. However, the comparison of the results for calculation enthalpies of IL synthesis reactions from G3MP2 and CBS-QB3 methods given in Table 4.21 shows an acceptable agreement within 2-3 $kJ \cdot mol^{-1}$ (except for [C$_1$mim][Cl]).

Table 4.21: Comparison of the results for calculation $\Delta_r H^0_{g,IL}$ from G3MP2 and CBS $kJ \cdot mol^{-1}$

IL	G3MP2	CBS-QB3
[C$_1$mim][Cl]	-3.1	-11.0
[C$_2$mim][Cl]	-6.8	-14.1
[C$_3$mim][Cl]	-10.7	-14.9
[C$_4$mim][Cl]	-12.0	-15.2
[C$_5$mim][Cl]	-12.5	-15.3
[C$_6$mim][Cl]	-12.7	-11.0

It is also necessary to mention that the G3MP2 method is not parameterized for calculation of the Br-containing molecules. That is why we used CBS-QB3 systematically throughout this work.

4.5 DSC - conclusion

There are several very important advantages of using DSC technique for the indirect evaluation of the vaporization enthalpies of ionic liquids:

- $\Delta_r H^0_{liq,IL}$ measurements using DSC are less demanding comparable to the combustion calorimetry;

- purity requirements for the chemicals are less rigorous comparable to the combustion calorimetry;

- $\Delta_r H^0_{g,IL}$ obtained using first principles ***directly*** from the calculated H_{298} (avoiding usual calculation of the enthalpies of formation with help of atomization or isodesmic procedures;

- $\Delta_{vap}H$ of the starting chemicals are well known or easy to be measured;

- DSC calorimeter is available as basic equipment of the laboratories working with ILs.

DSC standard equipment have been applied for the express determination of the reaction enthalpies of the ionic liquids synthesis reactions. The combination of the DSC method with the high-level first-principles calculations produces reliable sets of molar enthalpies of vaporization and enthalpies of formation for the series of the imidazolium, pyridinium, and pyrrolidinium based ILs. Simple additive rules have been tested for these thermochemical properties. We have shown that enthalpies of formation and enthalpies of vaporization generally obey group additivity and the values of the additivity parameters for ionic liquids seem to be not much different from those for molecular compounds. Combination of differential scanning calorimetry with the *first-principles* calculations allows rapid accumulation of the thermodynamic data. This procedure provides indispensable data material for testing first-principles procedures and molecular dynamic simulations techniques in order to understand thermodynamic properties of ionic liquids on a molecular basis.

5 Outlook and conclusion of the work

Having established the express experimental TGA procedure, an accumulation of vaporization enthalpies of ILs of different structures could be possible. Two directions of investigation could be suggested. The first one is to investigate series of ILs with the same anion and 1-Me-Imidazolium based cations with different alkyl chain length [$C_n$mim]. The second direction is to keep constant cation (e.g. [C_2mim]) and to change anion. Following these two directions it is possible to investigate a large scope of structure-property relationship data.

TGA method being a screening tool allows to locate values of $\Delta_{vap}H$ and to select proper temperature range of investigation with other well established techniques (static, Knudsen QCM). Simultaneous measurements with several methods give an opportunity to establish reliable vaporization enthalpy data for further applications.

At the moment reproducibility of the TGA method lies in the range of $\pm(3$ to $5)$ $kJ \cdot mol^{-1}$ advantages: it is easily accessible, fast and requires small amounts of sample. This is why TGA is a prospective method for $\Delta_{vap}H$ studies and needs further development.

For the reason of establishment of mutual consistency of $\Delta_{vap}H$ data other, indirect methods are also required. The procedure in which vaporization enthalpy can be determined was suggested. In this procedure $\Delta_{vap}H$ is obtained as a difference between enthalpies of formation of an IL in liquid and gaseous phase. Enthalpy of formation in gaseous phase can be computed using high level first principles calculation. Enthalpy of formation in liquid phase can be measured experimentally either with the use of combustion calorimetry or with the use of DSC.

Establishment of DSC for $\Delta_{vap}H$ studies is possible due to diversity of advantages of DSC, e.g.: the measurement of enthalpy of formation is less demanding comparing with combustion calorimetry (purity of reactants, time requirements, weighing procedure), simplicity and accessibility of DSC equipment. Being applied to $\Delta_{vap}H$ studies, DSC provides us with possibility to derive $\Delta_{vap}H$ easily from enthalpy of formation of the certain IL with high accuracy. Excessive data on thermodynamic properties of ILs obtained with the use of this indirect procedure give an opportunity to adjust parameters of *ab initio* and MD-simulations. In one turn, these simulations render essential data on structure-property relations for ILs, which allow to tune properties of these compounds more precisely.

References

[1] Hiroyuki Ohno, *Bull Chem. Soc. Jpn.*, **2006**, 75(11), 1665-1680.

[2] Thomas Welton, August , **1999**, *99(8)*, 2071-2084.

[3] Colin F. Poole, May , **2004**, *1037(1-2)*, 49-82.

[4] P. Wasserscheid and W. Keim, *Angew. Chem. Int. Ed.*, **2000**, *39*, 3772.

[5] D. Zhao, Y. Liao, and Z. Zhang, *Clean*, **2007**, *35,* 42.

[6] R.P. Swatloski, J.D. Holbrey, and R.D. Rogers, *Green Chem.*, **2003**, *5,* 361.

[7] Ch. P. Fredlake, J. M. Crosthwaite, D. G. Hert, S. N. V. K. Aki, and J. F. Brennecke, *J. Chem. Eng. Data*, **2004**, *49,* 954-964.

[8] H. C. Joglekar, I. Rahman, and B. D. Kulkarni, *Chem. Eng. Technol*, **2007**, *30,* 819.

[9] D. Rooney, J. Jacquemin, and R. Gardas, *Top. Curr. Chem.*, **2010**, *290,* 185.

[10] R. D. Rogers and K. R. Seddon, *Science,* **2003**, *302,* 792.

[11] S. Aparicio, M. Atilhan, and F. Karadas, *Ind. Eng. Chem. Res.*, **2010**, *49,* 95809595.

[12] L.P.N. Rebelo, J.N. Canongia Lopes, J.M.S.S. Esperanca, and E. Filipe, *J. Phys. Chem. B,* **2005**, *109,* 6040-6043.

[13] Y.U. Paulechka, Dz. H. Zaitsau, G.J. Kabo, and A.A. Strechan, *Thermochimica Acta,* **2005**, *439,* 158-160.

[14] M. J. Earle, J.M.S.S. Esperanca, M.A. Gilea, J.N. Canongia Lopes, L.P.N. Rebelo, J.W. Magee, K.R. Seddon, and J.A. Widegren, *Nature,* **2006**, *439,* 831-834.

[15] Th. Koeddermann, D. Paschek, and R. Ludwig, *Phys. Chem. Chem. Phys.*, **2008**, *9,* 549-555.

[16] R. Ludwig and U. Kragl, *Angew. Chem. Int. Ed.*, **2007**, *46,* 6582.

[17] J. M. S. S. Esperanca, J. N. Canongia, M. Tariq, L. M. N. B. F. Santos, J. W. Magee, and L. P. N. Rebelo, *J. Chem. Eng. Data*, **2010**, *55,* 3-12.

[18] R.D. Weir and Th. W. de Loos, *Measurement of Thermodynamic Properties of Multiple Phases;* Elsevier, Amsterdam, **2005**.

[19] J.L. Margrave, *The Characterization of High-Temperature Vapours;* Wiley, New York, **1967**.

[20] A.V. Suvorov, *Thermodynamic Chemistry of Vapor State;* Chemistry, Leningrad, **1970**.

[21] H. Kasheghari, I. Mokbel, C. Viton, and J. Jose, *Fluid Phase Equilibria*, **1993**, *87*, 133.

[22] W. Swietoslawski, *Ebulliometric measurements;* Reinhold, New York, **1945**.

[23] E. Hala, J. Piek, V. Fried, and O. Vilim, *Vapour Liquid Equilibium;* Pergamon Press, Oxford, **1967**.

[24] M. Rogalski and S. Malanowski, *Fluid Phase Equilibria*, **1980**, *5*, 97.

[25] S. Malanowski, *Fluid Phase Equilibria*, **1982**, *9*, 197-219.

[26] D. Ambrose, M.B. Ewing, N.B. Ghiassee, and J.C. Sanchez Ochoa, *J. Chem. Thermodyn.*, **1990**, *22*, 589.

[27] H.V. Reagnault, *Ann. Chim. (Paris)*, **1854**, *15*, 129.

[28] A.N. Nesmeyanov, *Vapor Pressure of the Chemical Elements;* Elsevier, Amsterdam, **1963**.

[29] R. Pankajavalli, C. Mallika, O.M. Sreedharan, V.S. Raghunthan, Antony Premkumar P., and K.S. Nagaraja, *Chem. Ing. Sei.*, **2002**, *57*, 3603.

[30] R. Pankajavalli, K. Ananthasivan, S. Antonysamy, and P.R. Vasudeva Rao, *J. Nuc. Mat*, **2005**, *345*, 96.

[31] I. Langmuir, *Phys. Rev.*, **1913**, *2*, 329.

[32] D.M. Price and M. Hawkins, *Thermochim. Acta*, **1998**, *315*, 19-24.

[33] D.M. Price and M. Hawkins, *Thermochim. Acta*, **1999**, *329*, 73.

[34] D.M. Price, *J. Therm. Anal. Cal*, **2001**, *64*, 315.

[35] Lori Burnham, David Dollimore, and Kenneth Alexander, *Thermochim. Acta*, **2001**, *367-368*, 15-22.

[36] Koustuv Chatterjee, Anasuya Hazra, David Dollimore, and S. Alexander Kenneth, *Journal of Pharmaceutical Sciences*, **2002**, *91*, 1156-1168.

[37] Koustuv Chatterjee, Anasuya Hazra, David Dollimore, and S. Alexander Kenneth, *European Journal of Pharmaceutics and Biopharmaceutics*, **2002**, *54*, 171180.

[38] W. Guekel, G. Synnatsehke, and R. Rittig, *Pestic. Sei.*, **1973**, *4*, 137-147.

[39] R. M. Stephenson and S. Malanowski, *Handbook of the Thermodynamics of Organic Compounds;* New York, Elsevier, **1987**.

[40] T. Boublik, V. Fried, and E. Hala, *The Vapor Pressure of Pure Substances;* Amsterdam, Elsevier, **1984**.

[41] P.G. Wahlbeek, *High. Temp. Set.*, **1986**, *21*, 189.

[42] J.M. Lafferty, Eds., *Scientific Foundations of Vacuum Technique;* Wiley, New York, **1962**.

[43] E.M. Eusebio, A. J. L. Jesus, M. S. C. Cruz, M. L. P. Leitão, and J. S. Redinha, J. Chem. Thermod., **2003**, *35*, 123.

[44] J.F. Liebman and A. Greenberg, Ed., *Molecular Structure and Energetics*, vol. 2; VCH Publisher, New York, **1987**.

[45] S.F. Donovan, J. *Chromatogr. A*, **1996**, *749*, 123.

[46] J.S. Chiekos, S. Hosseini, and D.G. Hesse, *Thermochim. Acta*, **1995**, *249*, 41.

[47] T. Kiyobayashi and M.E. Minas de Piedade, J. *Chem. Thermodyn.*, **2001**, *33*, 11.

[48] Federica Barotini and Valerio Cozzani, *Thermochim. Acta*, **2007**, *460*, 15-21.

[49] M. Knudsen, *The Kinetic Theory of Gases;* London, Methuen, **1950**.

[50] CG. De Kruif, T. Kuipers, J.C. Van Miltenburg, R.C.F. Sehaake, and G. Stevens, J. *Chem. Thermodyn.*, **1981**, *13*, 1081-1086.

[51] J.P. Elder, J. *Therm. Anal. Cal*, **1997**, *49*, 897-905.

[52] Koustuv Chatterjee, David Dollimore, and S. Alexander Kenneth, *Int. J. Pharm.*, **2001**, *213*, 31-44.

[53] Koustuv Chatterjee, David Dollimore, and S. Alexander Kenneth, J. *Therm. Anal*, **2001**, *63*, 629-639.

[54] P. Phang and D. Dollimore, *Proc. NATAS Annu. Conf. Therm. Anal Appl*, **1999**, *27*, 598-601.

[55] P. Phang, D. Dollimore, and S. J. Evans, *Proc. NATAS Annu. Conf. Therm. Anal Appl*, **2000**, *28*, 54-59.

[56] Lori Burnham, David Dollimore, and Kenneth Alexander, *Proc. NATAS Annu. Conf Therm. Anal Appl*, **1999**, *27*, 602-606.

[57] S. F. Wright, Kenneth Alexander, and David Dollimore, *Thermochim. Acta*, **2001**, *367*, 29-35.

[58] Koustuv Chatterjee, David Dollimore, and S. Alexander Kenneth, *Instr. Sci. Technol*, **2001**, *29*, 133-144.

[59] D.M. Price, *Thermochim. Acta*, **2001**, *376*, 367-368.

[60] E. A. Moelwyn-Hughes, *Physical chemistry;* Pergamon Press, London, **1957**.

[61] E. A. Mortensen and H. Eyring, J. *Phys. Chem.*, **1960**, *64*, 846.

[62] J. R. Maa, *Ind. Eng. Chem. Fundamen.*, **1967**, *6 (4)*, 504-518.

[63] W. W. Focke, J. *Therm. Anal Cal*, **2003**, *74*, 97-107.

[64] Koustuv Chatterjee, David Dollimore, and Kenneth S. Alexander, **2002**, *392-393*, 107-117.

[65] Stefano Veeehio, J. *Therm. Anal Cal,* **2006**, *84 (1),* 271-278.

[66] Supaporn Lerdkanchanaporn and David Dollimore, December, **1998**, *324(1-2),* 15-23.

[67] Niel Pieterse and Walter W. Focke, *Thermochim. Acta,* **2003**, *406,* 191-198.

[68] Stefano Vecchio, J. *Therm. Anal Cal,* **2007**, *87 (1),* 79-83.

[69] Stefano Vecchio and Mauro Tomassetti, *Fluid Phase Equilibria,* **2009**, *279,* 6472.

[70] Stefano Vecchio and Bruno Brunetti, J. *Chem. Thermodynamics,* **2009**, *41,* 880667.

[71] P. James Armstrong, Christopher Hurst, G. Robert Jones, Peter Licence, R. J. Kevin Lovelock, Satterley J. Christopher, and Ignacio J. Villar-Garcia, *Phys. Chem. Chem. Phys.,* **2007**, *9,* 982-990.

[72] Dzmitry H. Zaitsau, Gennady J. Kabo, Aliaksei A. Strechan, Yauheni U. Paulechka, Anna Tschersich, Sergey P. Verevkin, and Andreas Heintz, J. *Phys. Chem. A,* **2006**, *110,* 7303-7306.

[73] V. N. Emel'yanenko, S. P. Verevkin, and A. Heintz, J. *Am. Chem. Soc,* **2007**, *129,* 3930-3937.

[74] A. Deyko, K.R. Lovelock, P. Licence, and R.G. Jones, *Phys. Chem. Chem. Phys.,* **2009**, *11*. 8544-8555.

[75] K.R. Lovelock, A. Deyko, P. Licence, and R.G. Jones, *Phys. Chem. Chem. Phys.,* **2010**, *12,* 8893-8901.

[76] Andreas Seeberger, Ann-Katharin Andersen, and Andreas Jess, *Phys. Chem. Chem. Phys.,* **2009**, *11,* 9375-9381.

[77] F. Heym, B. J. M. Etzold, Ch. Kerna, and A. Jess, *Phys. Chem. Chem. Phys.,* **2010**, *12,* 12089-12100.

[78] O. Aschenbrenner, S. Supasitmongkol, M. Taylor, and P. Styring, *Green Chem.,* **2009**, *11,* 1217-1221.

[79] J. S. Chickos and W. E. Jr. Acree, J. *Phys. Chem. Ref. Data,* **2002**, *31,* 537-698.

[80] J. S. Chickos and W. E. Acree, J. *Phys. Chem. Ref. Data,* **2003**, *32,* 519-878.

[81] V. Major and V. Svoboda, *Enthalpies of Vaporization of Organic Compounds: A Critical Review and Data Compilation;* Blackwell Scientific Publications, Oxford, **1985**.

[82] S. P. Verevkin, *Private communication,* **2009**.

[83] Zaitsau, *Private communication,* **2010**.

[84] E. C. W. Clarke and D. N. Glew, *Trans. Faraday Soc,* **1966**, *62,* 539-547.

[85] Huimin Luo, A. Gary Baker, and Sheng. Dai, J. *Phys. Chem. B.,* **2008**, *112,* 10077-10081.

[86] C. Wang, H. Luo, H. Li, and S. Dai, *Phys. Chem. Chem. Phys.*, **2010**, *12*, 7246–7250.

[87] M. Mansson, P. Sellers, G. Stridh, and S. Sunner, *J. Chem. Therm.*, **1977**, *9*, 91-97.

[88] L.M.N.B.F. Santos, J.N. Canongia Lopes, J.A.P. Coutinho, J.M.S.S. Esperanca, L.R. Gomes, I.M. Marrucho, and L.P.N. Rebelo, *J. Am. Chem. Soc.*, **2007**, *129*, 284-285.

[89] V. N. EmePyanenko, S.P. Verevkin, Heintz A., J-A. Corfield, A. Deyko, K.R.J. Lovelock, P. Licence, and G.R. Jones, *J. Phys. Chem. A*, **2008**, page submitted.

[90] R. G. Jones, A. Armstrong, J. P. Deyko, C. Hurst, P. Licence, C. J. Satterley, K. R. Lovelock, A. Taylor, and I. J. Villa-Garcia, August , **2007**, pages 3P04-054.

[91] V. N. Emel'yanenko, S.P. Verevkin, Heintz A., and C. Schick, *J. Phys. Chem.A*, **2008**, page submitted.

[92] S.P. Verevkin, *J. Chem. Therm.*, **2006**, *38*, 1111-1123. [93] S.P.

Verevkin, *J. Chem. Therm.*, **2005**, *37*, 73-81.

[94] D. Kulikov, S.P. Verevkin, and Heintz A., *Fluid Phase Equilibria*, **2001**, *192*, 187-207.

[95] L. Bodo, L. Gontrani, R. Caminiti, N.V. Plechkova, K.S. Seddon, and Triolo. A., *J. Phys. Chem. B*, **2010**, *114*, 16398-16407.

[96] H.V.R. Harsha V. R. Annapureddy, H.K. Kashyap, P.M. De Biase, and C.J. Margulis, *J. Phys. Chem. B*, **2010**, *114*, 16838-16846.

[97] J.R. Stuff, *Thermochim. Acta*, **1989**, *152*, 421-425.

[98] Dzmitry H. Zaitsau, Yauheni U. Paulechka, and Gennady J. Kabo, *J. Phys. Chem. A*, **2006**, *110*, 11602-11604.

[99] R. E. del Sesto, T. M. McCleskey, A. K. Burrell, G. A. Baker, J. D. Thompson, L. Scott, J. S. Wilkes, and P. Williams, *Chem. Commun.*, **2008**, pages 447449.

[100] T. Sasaki, C. Zhong, M. Tada, and Y. Iwasawa, *Chem. Commun.*, **2005**, pages 2506-2508.

[101] S. Hayashi and H. Hamaguchi, *Chem. Lett*, **2004**, *33*, 1590-1591.

[102] S. Hayashi, S. Saha, and H. Hamaguchi, *IEEE Trans. Magn.*, **2006**, *42*, 12-14.

[103] Q.-G. Zhang, J.-Z. Yang, X.-M. Lu, J.-S. Gui, and M. Huang, *Fluid Phase Equilibria*, **2004**, *226*, 207-211.

[104] Gjergj Dodbiba, Hyun Seo Park, Katsunori Okaya, and Toyohisa Fujita, *J. of Magn. and Magn. Mater.*, **2008**, *320(7)*, 1322-1327.

[105] C. Guerrero-Sanchez, T. Lara-Ceniceros, E. Jimenez-Regalado, M. Rasa, and U.S. Schubert, *Adv. Mater.*, **2007**, *19(13)*, 1740-1747.

[106] Rico E. Del Sesto, Cynthia Corley, Al Robertson, and John S. Wilkes, *Journal of Organometallic Chemistry*, **2005**, *690(10)*, 2536-2542, 16th International Conference on Phosphorus Chemistry, International Conference on Phossphorus Chemistry.

[107] Oleg Borodin, **2009**, *113(36)*, 12353-12357.

[108] A. Renken, V. Hessel, P. Löb, R. Miszczuk, M. Uerdingen, and L. Kiwi-Minsker, *Chemical Engineering and Processing*, **2007**, *46(9)*, 840-845, Selected Papers from the European Process Intensification Conference (EPIC), Copenhagen, Denmark, September 19-20, **2007**.

[109] G. Hoehne, W. Hemminger, and H.-J. Flammersheim, *Differential scanning calorimetry;* Springer-Verlag, Berlin-Heidelberg, **2003**.

[110] Donna G. Blackmond, Thorsten Rosner, and Andreas Pfaltz, July, **1999**, *3(A)*, 275-280.

[111] H. Loewe, R.D. Axinte, D. Breuch, and C. Hofmann, December, **2009**, *155(1-2)*, 548-550.

[112] Alexandra Grosse Boewing and Andreas Jess, March, **2007**, *62(6)*, 1760-1769.

[113] Alexandra Grosse Boewing and Andreas Jess, *Green Chem.*, **2005**, *7(4)*, 230-235.

[114] Jay C. Schleicher and Aaron M. Scurto, *Green Chem.*, **2009**, *11 (5)*, 694-703.

[115] P. Wasserscheid and W. Keim, *Angewandte Chemie International Edition*, **2000**, *39(21)*, 3772-3789.

[116] John R. Stuff, October, **1989**, *152(2)*, 421-425.

[117] Helen L. Ngo, Karen LeCompte, Liesl Hargens, and Alan B. McEwen, **2000**, *357-358*, 97-102.

[118] Ibrahim Shehatta, January, **1993**, *213*, 1-10.

[119] John D. Holbrey, Reichert W. Matthew, Reddy Ramana G., and Rogers Robin D.; Heat capacities of ionic liquids and their applications as thermal fluids; In *ACS Symposium Series*, Vol. 856, pages 121 133 . American Chemical Society, **2003**.

[120] Yauheni U. Paulechka, **2010**, *39(3)*, 033108-23.

[121] Yauheni U. Paulechka, Audrey G. Kabo, and Audrey V. Blokhin, **2009**, *113(AA)*, 14742-14746.

[122] Y.U. Paulechka, G.J. Kabo, A.V. Blokhin, A.S. Shaplov, E.I. Lozinskaya, and Ya.S. Vygodskii, **2007**, *39(1)*, 158-166.

[123] Gennady J. Kabo, Yauheni U. Paulechka, Audrey G. Kabo, and Andrey V. Blokhin, **2010**, *42(10)*, 1292-1297.

[124] M. J. Frisch, G. W. Trucks, and H. B. Schlegel, *Gaussian 03, Revision B.04;* Gaussian, Inc., Pittsburgh, PA, **2003**.

[125] J. A. Montgomery, Jr., M. J. Frisch, J. W. Ochterski, and G. A. Petersson, April , **2000**, *112(15),* 6532-6542.

[126] D.A. McQuarrie, *Statistical Mechanics;* Harper & Row, New York, **1976**.

[127] A. Gross Boewing; *Zur Kinetik und Reaktionstechnik der Synthese ionischer Fluessigkeiten;* PhD thesis, Uni-Bayreuth, **2006**.

[128] A. Jess, A. Große Böwing, and P. Wasserscheid, *Chemie Ingenieur Technik,* **2005**, 77(9), 1430-1439.

[129] Daniel A. Waterkamp, Michael Heiland, Michael Schlüter, Janelle C. Sauvageau, Tom Beyersdorff, and Jorg Thoming, *Green Chem.,* **2007**, *9(10),* 1084-1090.

[130] Shaozheng Hu, Anjie Wang, Holger Löwe, Xiang Li, Yao Wang, Changhong Li, and Dong Yang, **2010**, *162(1),* 350-354.

[131] H. Loewe, R.D. Axinte, D. Breuch, C. Hofmann, J.H. Petersen, R. Pommersheim, and A. Wang, **2010**, *163(3),* 429-437.

[132] John D. Holbrey, W. Matthew Reichert, Richard P. Swatloski, Grant A. Broker, William R. Pitner, Kenneth R. Seddon, and Robin D. Rogers, *Green Chem.,* **2002**, *4(5),* 407-413.

[133] W.N. Hubbard, D.W. Scott, and G. Waddington, *Experimental Thermochemistry;* Interscience, New York, **1956**.

[134] Clemens B. Minnich, Lukas Küpper, Marcel A. Liauw, and Lasse Greiner, **2007**, *126(1-2),* 191-195.

[135] V. N. Emel'yanenko, S.P. Verevkin, Heintz A., and C. Schick, July , **2008**, *112(27),* 8095-8098.

[136] Vladimir N. Emel'yanenko, Sergey P. Verevkin, Andreas Heintz, Karsten Voss, and Axel Schulz, **2009**, *113(29),* 9871-9876.

[137] Sergey P. Verevkin, Vladimir N. Emel'yanenko, Dzmitry H. Zaitsau, Andreas Heintz, Chris D. Muzny, and Michael Frenkel, *Phys. Chem. Chem. Phys.,* **2010**, *12(45),* 14994-15000.

[138] Dzmitry H. Zaitsau, Vladimir N. Emel'yanenko, Sergey P. Verevkin, and Andreas Heintz, **2010**, 55(12), 5896-5899.

[139] Otilia Mo, Manuel Yanez, Maria Victoria Roux, Pilar Jimenez, Juan Z. Davalos, Manuel A. V. Ribeiro da Silva, Maria das Dores M. C. Ribeiro da Silva, M. Agostinha R. Matos, Luisa M. P. F.

Amaral, Ana Sanchez-Migallon, Pilar Cabildo, Rosa Claramunt, Jose Elguero, and Joel F. Liebman, **1999**, *103(46)*, 9336-9344.

[140] G. Stridh and S. Sunner, **1975**, 7(2), 161-168.

[141] L. Bjellerup, *Acta Chem. Scand.*, **1961**, *15*, 231-241.

[142] H. Mackle and W.V. Steele, *Trans. Faraday Soc*, **1969**, *65*, 2053-2059.

[143] W. N. Hubbard, F. R. Frow, and Guy Waddington, August, **1961**, *65(8)*, 1326–1328.

[144] V. N. Emel'yanenko, S.P. Verevkin, Heintz A., J-A. Corfield, A. Deyko, K. Lovelock, P. Licence, and G.R. Jones, September, **2008**, *112(37)*, 11734-11742.

[145] Kevin R. J. Lovelock, Alexey Deyko, Jo-Anne Corfield, Peter N. Gooden, Peter Licence, and Robert G. Jones, *Chem. Eur. J. of Chem. Phys.*, **2009**, *10(2)*, 337-340.

[146] Steven D. Chambreau, Ghanshyam L. Vaghjiani, Albert To, Christine Koh, Daniel Strasser, Oleg Kostko, and Stephen R. Leone, January, **2010**, *114(3)*, 1361-1367.

[147] S. P. Verevkin, *Angewandte Chemie*, **2008**, *47*, 5071-5074.

[148] W. S. Ohlinger, P, E, Klunzinger, B, J, Deppmeier, and W. J. Hehre, March, **2009**, *113(10)*, 2165-2175.

[149] S. P. Verevkin, V. N. Emel'yanenko, Pimerzin A. A., and Vishnevskaya E. E., *J. Phys. Chem. A*, **2011**, page in press.

[150] Ingemar Wadsoe, *Acta Chem. Scand.*, **1968**, *22*, 2438-2444.

[151] Mati Karelson, Victor S. Lobanov, and Alan R. Katritzky, *Chem. Rev.*, **1996**, *96(3)*, 1027-1044.

[152] Alan R. Katritzky, Ritu Jain, Andre Lomaka, Ruslan Petrukhin, Mati Karelson, Ann E. Visser, and Robin D. Rogers, *J. Chem. Inf. Comput. Sci.*, **2002**, *42(2)*, 225-231.

[153] Sydney W. Benson, *Thermochemical kinetics: methods for the estimation of thermochemical data and rate parameters;* Wiley, **1976**.

[154] N. Cohen, *J. Phys. Chem. Ref. Data*, **1996**, *25(6)*, 1411-1481.

[155] John M. Slattery, Corinne Daguenet, Paul J. Dyson, Thomas S. Schubert, and Ingo Krossing, *Angewandte Chemie*, **2007**, *119(28)*, 5480-5484.

[156] Eugen L. Krasnykh, Sergey P. Verevkin, Bohumir Koutek, and Jan Doubsky, *J. Chem. Therm.*, **2006**, *38(6)*, 717-723.

[157] Vladimir N. Emel'yanenko, Sergey P. Verevkin, Bohumir Koutek, and Jan Doubsky, *J. Chem. Therm.*, **2005**, 57(1), 73-81.

[158] J.P. Pedley, R.D. Naylor, and S.P. Kirby, *Thermochemical Data of Organic Compounds;* Ed. Chapmen and Hall, London, **1986**.

[159] Dmitry Kulikov, Sergey P. Verevkin, and Andreas Heintz, *Fluid phase equilibria*, **2001**, *192(1-2)*, 187-207.

[160] Sergey P. Verevkin, Eugen L. Krasnykh, Tatiana V. Vasiltsova, Bohumir Koutek, Jan Doubsky, and Andreas Heintz, *Fluid phase equilibria*, **2003**, *206(1-2)*, 331-339.

[161] Sergey P. Verevkin, *J. Chem. Eng. Data*, **2002**, *47(5)*, 1071-1097.

[162] R. Notario, O. Castano, J.-L. M. Abboud, R. Gomperts, L. M. Frutos, and R. Palmeiro, *J. Org. Chem.*, **1999**, *64(25)*, 9011-9014.

[163] Krishnan Raghavachari, Boris B. Stefanov, and Larry A. Curtiss, *J. Chem. Phys.*, **1997**, *106(16)*, 6764-6767.

[164] Herbert R. Kemme and Saul I. Kreps, *J. Chem. Eng. Data*, **1969**, *14(1)*, 98-102.

Acknowledgments

In this section I would like to express my deep acknowledge to people, who supported creation of this work.

On the first hand I would like to thank Prof. Dr. Sergej Verevkin for his guidance, ideas and remarkable support, which he gave me during his supervision. He encouraged me all the time to go straight to the aim and to conduct a lot of interesting experiments. Also I thank Prof. Dr. Christoph Schick for a nice working atmosphere in his group, for interesting discussions and ideas. I deeply appreciate his availability for discussions of any questions any time.

This work could not be possible without experiments conducted by Dr. Dzmitry Zaitsau. His experimental results were of the greatest importance in interpretation of my own measurements.

Quantum-chemical computations were also valuable. And I show my gratitude to Dr. Vladimir YemePyanenko for his efforts in such computations.

Additional acknowledgment should be addressed to Mr. Evgeny Shoifet for his advice how to apply LaTeX to creation of this manuscript.

All the members of the Polymer physics working group are acknowledged as well for the help in dealing with experimental equipment.

Cordial thanks are addressed to my family and my friends for their moral support in many situations.

Die VDM Verlagsservicegesellschaft sucht für wissenschaftliche Verlage abgeschlossene und herausragende

Dissertationen, Habilitationen, Diplomarbeiten, Master Theses, Magisterarbeiten usw.

für die kostenlose Publikation als Fachbuch.

Sie verfügen über eine Arbeit, die hohen inhaltlichen und formalen Ansprüchen genügt, und haben Interesse an einer honorarvergüteten Publikation?

Dann senden Sie bitte erste Informationen über sich und Ihre Arbeit per Email an *info@vdm-vsg.de*.

Sie erhalten kurzfristig unser Feedback!

VDM Verlagsservicegesellschaft mbH
Dudweiler Landstr. 99
D - 66123 Saarbrücken

Telefon +49 681 3720 174
Fax   +49 681 3720 1749

www.vdm-vsg.de

Die VDM Verlagsservicegesellschaft mbH vertritt

Printed by Books on Demand GmbH, Norderstedt / Germany